OXFORD CHEMISTRY PRIMERS

Physical Chemistry Editor
RICHARD G. COMPTON
Physical and Theoretical
Chemistry Laboratory
University of Oxford

Founding Editor and Organic
Chemistry Editor
STEPHEN G. DAVIES
The Dyson Perrins Laboratory
University of Oxford

Inorganic Chemistry Editor
JOHN EVANS
Department of Chemistry
University of Southampton

Chemical Engineering Editor
LYNN F. GLADDEN
Department of Chemical Engineering
University of Cambridge

Core Carbonyl Chemistry

John Jones

The Dyson Perrins Laboratory and Balliol College, University of Oxford

Series sponsor: **ZENECA**

ZENECA is a major international company active in four main areas of business: Pharmaceuticals, Agrochemicals and Seeds, Specialty Chemicals, and Biological Products.

ZENECA's skill and innovative ideas in organic chemistry and bioscience create products and services which improve the world's health, nutrition, environment, and quality of life. ZENECA is committed to the support of education in chemistry and chemical engineering.

OXFORD UNIVERSITY PRESS

This book has been printed digitally and produced in a standard specification
in order to ensure its continuing availability

OXFORD
UNIVERSITY PRESS

Great Clarendon Street, Oxford OX2 6DP

Oxford University Press is a department of the University of Oxford.
It furthers the University's objective of excellence in research, scholarship,
and education by publishing world-wide in

Oxford New York

Auckland Bangkok Buenos Aires Cape Town Chennai
Dar es Salaam Delhi Hong Kong Istanbul Karachi Kolkata
Kuala Lumpur Madrid Melbourne Mexico City Mumbai Nairobi
São Paulo Shanghai Taipei Tokyo Toronto

Oxford is a registered trade mark of Oxford University Press
in the UK and in certain other countries

Published in the United States
by Oxford University Press Inc., New York

ISBN 0-19-855959-3

Antony Rowe Ltd., Eastbourne

Series Editor's Foreword

The carbonyl group is the most common functional group in organic chemistry (aldehydes, ketones, acids, esters, amides etc) and therefore an understanding of its fundamental rectivity can be divided into just two categories—nucleophilic addition and α-hydrogen activity—it is surprising how baffling carbonyl chemistry can appear to beginners.

Oxford Chemistry Primers have been designed to provide concise introductions relevant to all students of chemistry and contain only the essential material that would be covered in an 3–10 lecture course. In Core Carbonyl Chemistry, John Jones provides a lucid and logical introduction to the basic chemistry of the carbonyl group—ideal for first year chemistry students and for later revision. the primer will be of interest to apprentice and master alike.

Professor Stephen G. Davies
The Dyson Perrins Laboratory
University of Oxford

Preface

An intimate understanding and instinctive feel for the chemistry of carbonyl compounds is absolutely essential for the modern organic chemist. The material set out here has been developed from a first-year lecture course which I have given for some years to students reading chemistry or biochemistry at Oxford, but here and there extends into what is conventionally deemed second-year territory. It owes more than my audiences will easily credit to the comment sheets handed in by them, which are sometimes as perceptive as they are barbed and irreverent. The lectures were of course not dreamt up by me out of the blue, but naturally follow the general direction of the path well-trodden by my forbears in the Dyson Perrins Laboratory. Of these, the late Sir Richard Norman, who lectured at Oxford when I was an undergraduate in the early sixties, and later with his great book (first edition 1968), probably shaped the way I think about general organic chemistry more than anyone else. One of the anonymous reviewers of the original proposal made a number of suggestions (cerium applications, SAMP, and Weinreb amides, for example), which at first struck me as a bit avant-garde. However, I warmed to them on reflection and adopted nearly all of them, so I must thank that reviewer. I am also grateful to my Balliol pupils Faisal Khan and Amber Haq for many useful comments. However the responsibility for the errors – whether of commission or omission – and the eccentricities of selection and arrangement, is all my own.

Oxford   J.H.J.
April 1997

Contents

Abbreviations

Ac	acetyl
aq.	aqueous
Ar	an aryl group
B	a Brønsted base*
Bu	*n*-butyl
Bu^i	*i*-butyl
Bu^t	*t*-butyl
cat. amt.	catalytic amount
DIBAL	diisobutylaluminium hydride
DME	dimethoxyethane
DMF	dimethylformamide
DMSO	dimethylsulfoxide
E or E$^+$	an electrophile
equiv.	equivalent
Et	ethyl
HA	a Brønsted acid
Hal	halogen (except F)
HMPA	hexamethylphosphoramide
LA	a Lewis acid
LDA	lithium diisopropylamide
Me	methyl
Nu or Nu$^-$	a nucleophile
Ph	phenyl
Pr^i	*i*-propyl
R	an alkyl group**
rds	rate-determining step
THF	tetrahydrofuran
X	a leaving group
xs.	excess
Δ	heat
~	the rest of the molecule

* Occasionally boron, but context will distinguish.
** Sometimes an alkyl *or* aryl group, *or* H; when general points about reactivity are being illustrated, the R symbols which might have been employed to complete partial structures are often omitted for the sake of clarity.

1 Introduction: the main themes

The carbonyl group is the commonest reactive component of organic structures. The most important functional groups containing it are given in the following list.

The common carbonyl-containing functional groups

Methanal or formaldehyde	![structure] $\underset{H \quad H}{\overset{O}{\parallel}}$
An aldehyde	$\underset{R \quad H}{\overset{O}{\parallel}}$
A ketone; the groups R may be joined	$\underset{R \quad R}{\overset{O}{\parallel}}$
A carboxylic acid	$\underset{R \quad OH}{\overset{O}{\parallel}}$
An ester: a lactone if the groups R are joined	$\underset{R \quad OR}{\overset{O}{\parallel}}$
A primary amide	$\underset{R \quad NH_2}{\overset{O}{\parallel}}$
An *N*-alkylamide or secondary amide: a lactam if the groups R are joined	$\underset{R \quad NHR}{\overset{O}{\parallel}}$
An *N,N*-dialkylamide or tertiary amide	$\underset{R \quad NR_2}{\overset{O}{\parallel}}$
An acid anhydride; the groups R may be joined	$\underset{R \quad O \quad R}{\overset{O \quad\quad O}{\parallel \quad\parallel}}$
An acyl or acid chloride	$\underset{R \quad Cl}{\overset{O}{\parallel}}$

But this list does not exhaust the possibilities, because these simple groups can be combined together in many ways, as exemplified in carbamates, imides and chloroformates.

Both an amide and an ester
—a urethane or carbamate ester

An imide; the groups R
may be joined

Both an acid chloride and an ester
—an alkoxycarbonyl chloride or a
chloroformate ester

From above

From the
side

Furthermore, sulfur can replace all or any of the oxygen atoms in all of these assemblies, and there are also the heterocumulenes to consider: $RR'C=C=O$ (ketenes), $RN=C=O$ (isocyanates), and $RN=C=S$ (isothiocyanates). We shall concern ourselves in this book mostly with aldehydes, ketones, and simple carboxylic acid derivatives, excluding nearly all the possible variations on these rich themes, even metal carbonyls and carbon monoxide itself. While passing over most members of this wildly diverse family, however, we hope some generally applicable principles can be illustrated.

In carbonyl compounds, both the carbon and oxygen atoms are sp^2-hybridized, with the σ-bonds to carbon and the oxygen lone pairs coplanar. The π-electron density above and below the plane is in a molecular orbital formed from the interaction of the p-orbital on carbon with that on oxygen. As the electronegativity of oxygen is greater than that of carbon, the carbonyl group is polarized, with the carbon atom positively charged relative to the oxygen, so that the carbon atom is electron-demanding. In resonance terms, there is a very significant contribution from a dipolar canonical structure. The magnitude of the partial positive charge on the carbonyl carbon also depends on the attached groups; it is decreased by electron-releasing groups, and increased by electron-withdrawing groups.

All the reactions of carbonyl compounds with which we shall be dealing involve electron supply to the electron-demanding carbon as a key feature. Such electron-supply can be achieved by either, (I) nucleophilic attack, or (II) ionization at a contiguous centre.

1.1 Electron-supply to the carbonyl carbon: (I) by nucleophilic attack

A nucleophile—let us take the anionic case Nu⁻ to begin with—can relieve the electron-demand of a carbonyl carbon by attack from above and behind, forming a bond to that atom, which rehybridizes from sp^2 to sp^3 as the reaction proceeds.

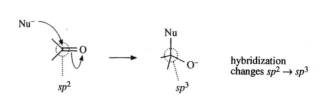

hybridization
changes $sp^2 \rightarrow sp^3$

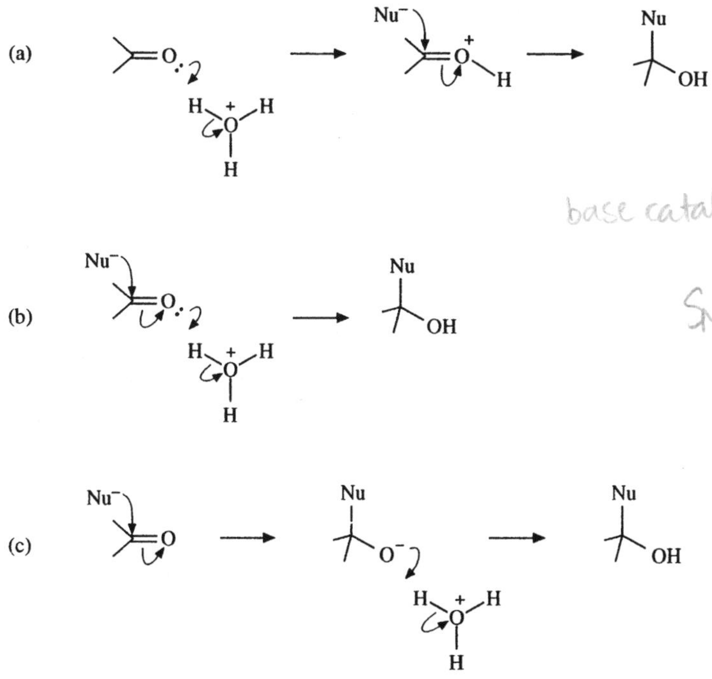

*acid
catalysis*

(a)

base catalysis

(b)

SN_1

(c)

SN_2

This process can be preceded by (a), simultaneous with (b), or followed by (c) interaction of an oxygen lone pair with a Lewis acid. The Lewis acid is commonly a proton, in which case the overall result is nucleophilic addition of NuH to the carbonyl double bond.

The nucleophile can also be electrostatically neutral, in which case the attacking atom tends to become positively charged, but the commonest scenario is one where the neutral nucleophile has a hydrogen atom which can depart, taking the charge with it.

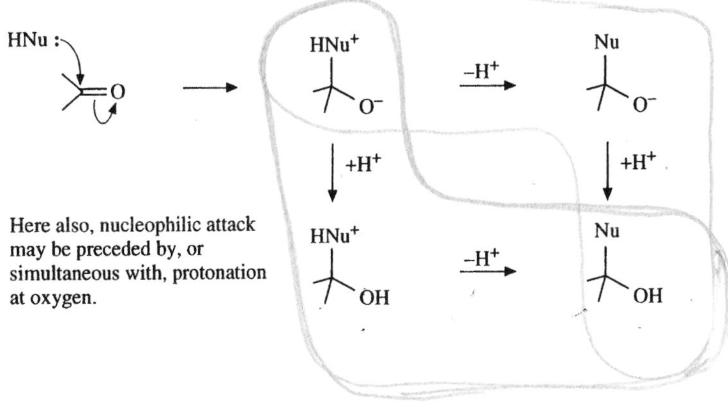

Here also, nucleophilic attack may be preceded by, or simultaneous with, protonation at oxygen.

In principle, albeit not always in practice, all these variations are reversible.

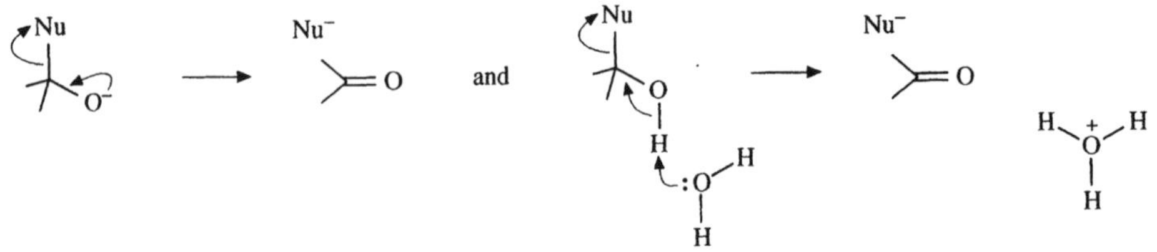

After nucleophilic addition there are, broadly speaking, six options open to the tetrahedral adduct, over and above collapse with regeneration of the reactants from which it was formed:

(1) no further reaction, e.g. hydration of aldehydes and ketones;

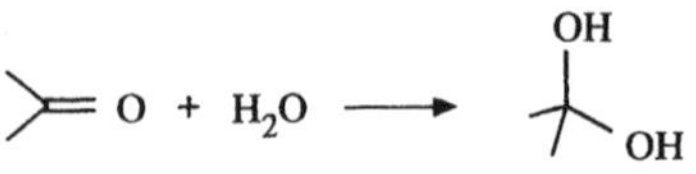

(2) the OH is replaced, e.g. acetal formation from aldehydes and ketones;

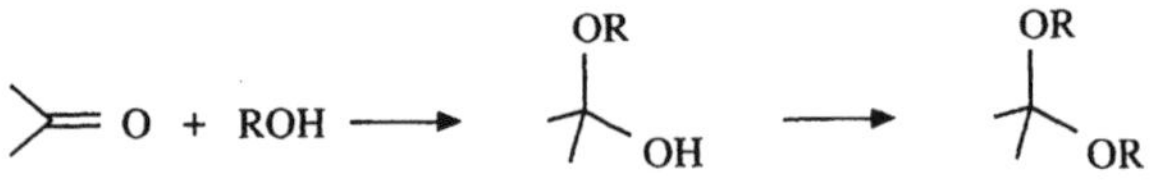

Hemiacetal   Acetal

(3) dehydration, i.e. overall condensation, e.g. reactions of aldehydes and ketones with amino compounds;

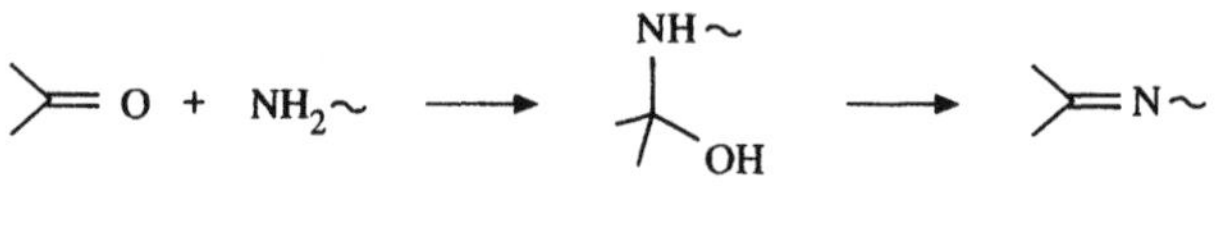

(with secondary amines, the adduct has no proton left on nitrogen to shed, but OH loss still takes place);

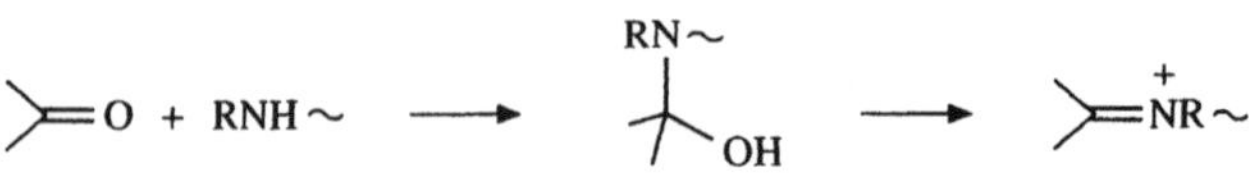

(4) loss of an attached group, i.e. overall substitution in carboxylic acid derivatives;

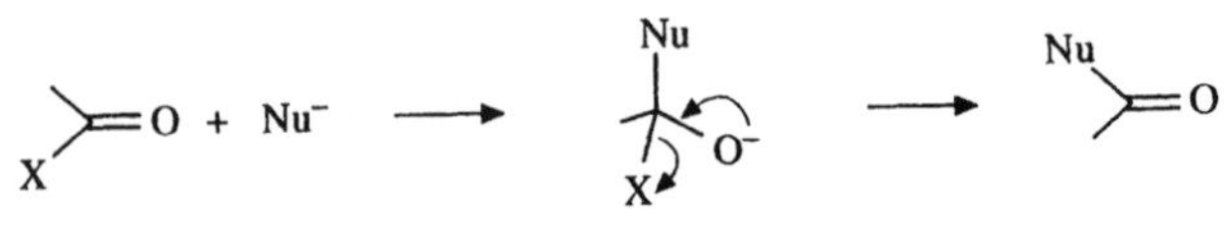

(5) skeletal rearrangement, if the adduct has an electrophilic site to which an electron-releasing attached group can migrate, e.g. the Baeyer–Villiger rearrangement;

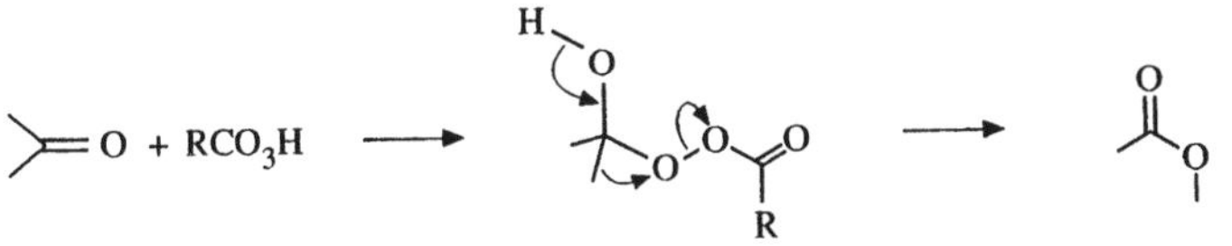

See Primer 5, p. 53

(6) the tetrahedral adduct oxygen may be trapped by an electrophilic site if one is brought within range by the nucleophilic reagent, e.g. the reaction of $Me_2S^+-CH_2^-$ with ketones.

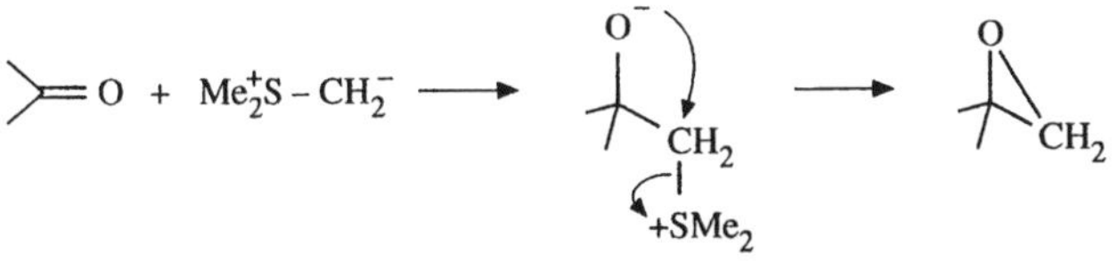

See Primer 33, p. 32

There are a vast number of variations on, and sequels to, options (1)–(4). Most of the rest of this book will be concerned with them. Analogues of option (5) are also numerous, but the treatment of molecular rearrangements is beyond our scope. Option (6) is rarely possible, but is further exemplified by the Stobbe and Darzens reactions (see Section 9.3).

1.2 Electron-supply to the carbonyl carbon: (II) by deprotonation of a contiguous atom

The ionization of carboxylic acids (pK_a 3–5, cf. 15–19 for alcohols) is a familiar process, driven by the delocalisation which is possible in the corresponding anion.

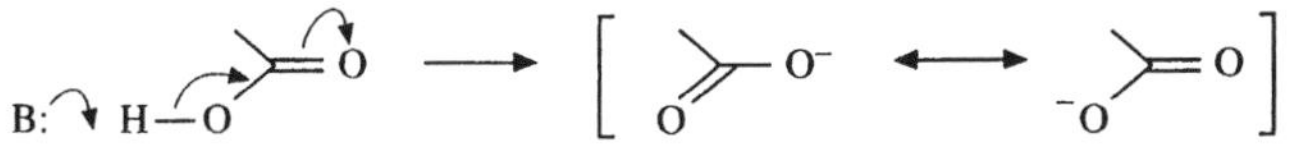

Carboxylate ion

Similarly, though much less easily, amides can ionize (pK_a 16–18, cf. 35 for amines).

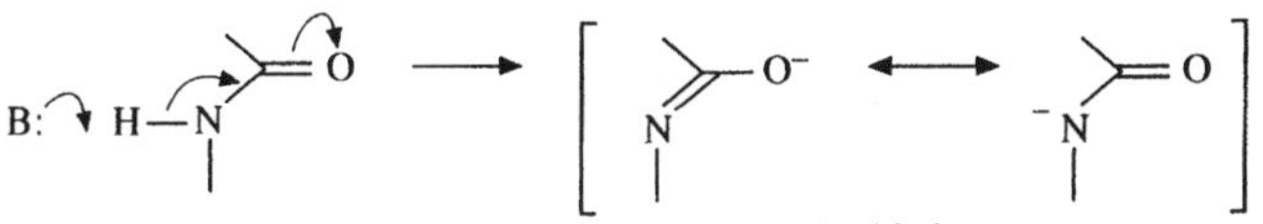

Amide ion

In the conjugate base of an imide, there is stabilization by delocalization of the negative charge over two carbonyl groups, so imides are more acidic than amides (imides pK_a 10–12, cf. amides 16–18).

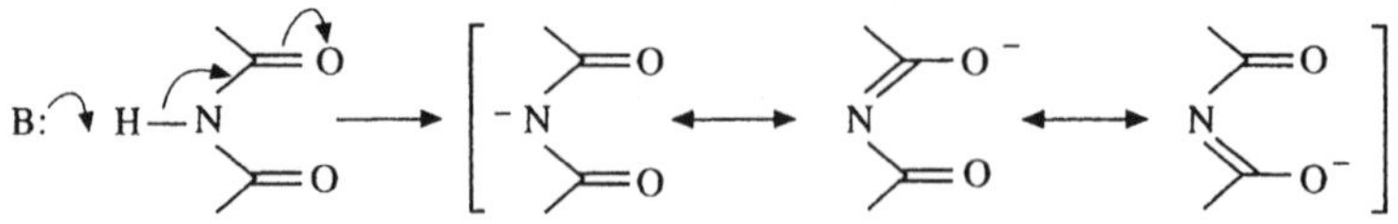

And α-CH is also acidic. Again, the more scope there is for delocalization, the more stable the anion and the more acidic the α-CH (pK_a CH$_3$COCH$_3$ 20, cf. CH$_3$COCH$_2$COCH$_3$ 9).

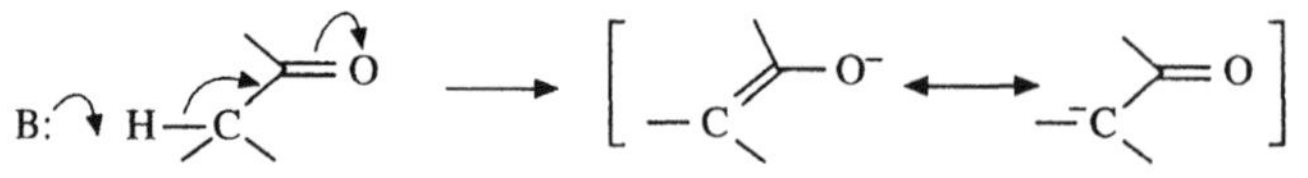

Enolate ion

The delocalized enolate ion produced by such ionization can reprotonate at carbon, giving the original carbonyl compound. Alternatively, it can protonate at oxygen, giving an isomeric structure called an enol (it is both an alk*ene* and an alcoh*ol*). The carbonyl (keto) structure and its enol isomer are tautomers (i.e. easily interconverted isomers), and the phenomenon of their interconversion is called keto–enol tautomerism.

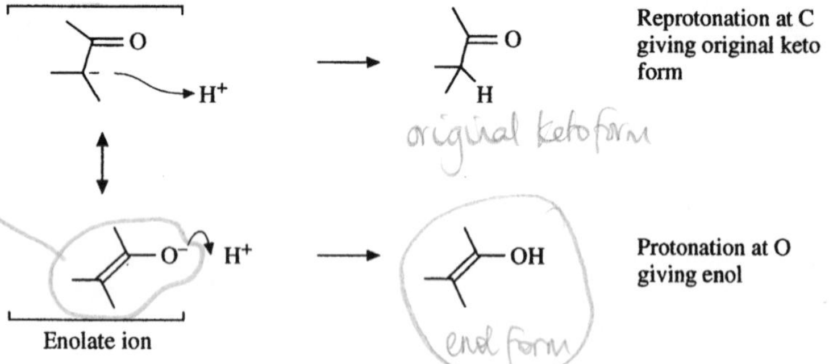

Reprotonation at C giving original keto form

original keto form

enolate

Enolate ion

Protonation at O giving enol

enol form

The enol can also be formed by *C*-deprotonation (a) simultaneous with, or (b) after *O*-protonation, without the intermediacy of an enolate ion.

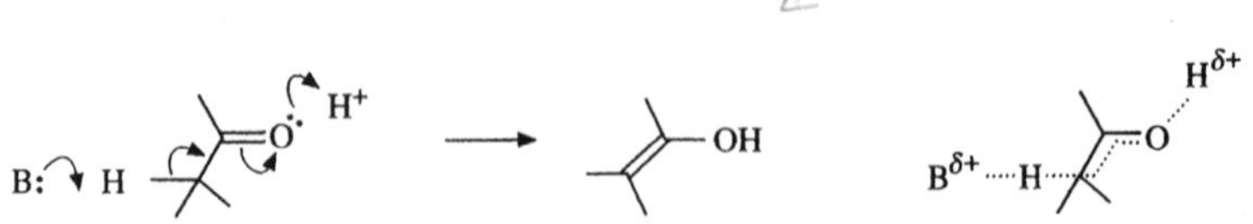

base catalysis — S$_N$2

(a) Concerted enol formation

Transition state

acid catalysis

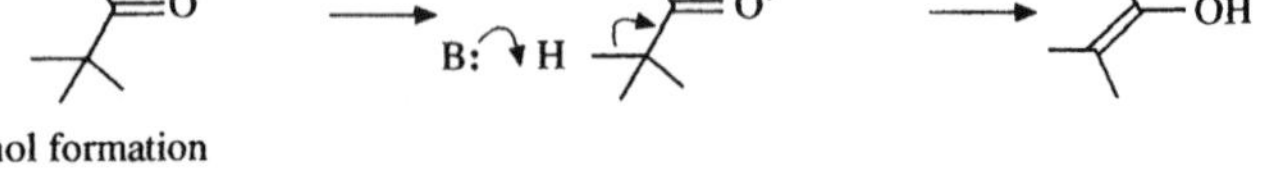

(b) Stepwise enol formation

(handwritten margin notes)

☒

how do you know whether a reaction will undergo a "concerted enol formation" or a "stepwise enol formation"?

These processes, involving protons jumping on and off all over the place, enable tautomeric equilibria between keto and enol forms to be established very easily, in all except completely neutral aprotic conditions. Such tautomeric equilibria are of considerable interest in themselves, and are of particular importance in organic synthesis, because enols and enolates can react usefully with a wide variety of electrophiles.

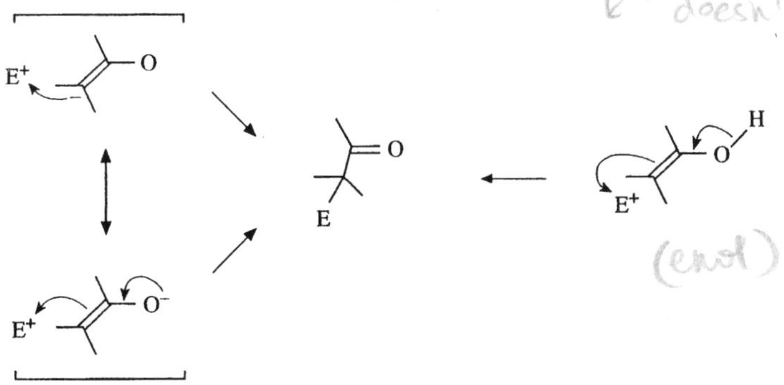

E⁺ doesn't have to be H⁺ all the time.

(enol)

Problems

1. Although methyl iodide reacts with simple enolates predominantly as shown above, O-methyl minor products can also be formed. How? Why are they minor products?

2. Acetyl chloride reacts very readily with NaOEt/EtOH, but 2-chloropropene is inert. Explain.

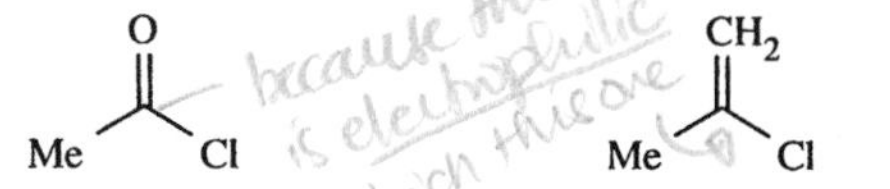

Acetyl chloride   2- Chloropropene

because this carbon is electrophilic which the one is not.

3. Cyclohexanone reacts with diazomethane (CH_2N_2) to give cyclo-heptanone and an isomeric epoxide. Explain.

> electrophilic Major resonance)
due to the (more stable
C=O polar bond, resonance)

minor (less stable resonance)

(?)

2 The addition of nucleophilic reagents to aldehydes and ketones

2.1 Structure and reactivity

H
H
Formaldehyde

100% hydrated

Me
H
Acetaldehyde

Me
Me
Acetone

100% unhydrated

Me
|
Me—CH
=O
Me—CH
|
Me

Diisopropyl ketone

In the series formaldehyde–acetaldehyde–acetone, reactivity to nucleophiles decreases in that order. This is partly because Me is electron-releasing compared to H, so successive H-replacement by Me diminishes the electrophilicity of the carbonyl carbon; and partly because Me is bigger than H, so H-replacement by Me introduces steric hindrance to approach by a nucleophile. If we go further with the introduction of Me groups, to, say, diisopropyl ketone, we find that this is much less reactive than acetone.

The electronic factor here is deactivating, but slight, because the additional Me groups are not attached directly to the carbonyl carbon. They are largely insulated from it by saturated carbon atoms. Unreactivity in this case is mainly down to steric effects, to which there are two contributions, both operating in the same sense. To begin with, obviously there is the sheer obstruction of the large isopropyl groups, making it difficult for a nucleophile to follow the necessary trajectory towards the electrophilic centre, which is from above and behind the carbonyl group. More subtle, in the trigonal ketone the two isopropyl groups are attached by bonds which diverge from each other at an angle of about 120°, but in the tetrahedral adduct this angle is reduced to about 109°: the additional internal compression between the bulky groups which results from nucleophilic addition and rehybridization militates against it.

An aryl group can release electron density by conjugation. Resonance

in this manner is not possible with alkyl groups, so aromatic aldehydes and ketones are less reactive towards nucleophiles than aliphatic aldehydes and ketones, respectively. Substituents on the aromatic ring can have a profound effect. Thus p-dimethylaminobenzaldehyde has a rather unreactive carbonyl group, because of electron-release by the dimethylamino nitrogen.

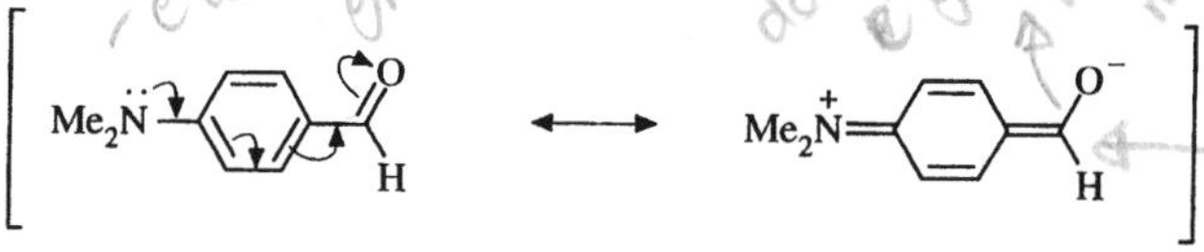

[handwritten: e⁻ donating group]
[handwritten: donating e⁻ group made the carbon more e⁻ rich]
[handwritten: this C is not very electrophilic anymore.]

A *p*-nitro substituent, on the other hand, pulls electrons away from the carbonyl group in *p*-nitrobenzaldehyde, which is consequently comparatively reactive.

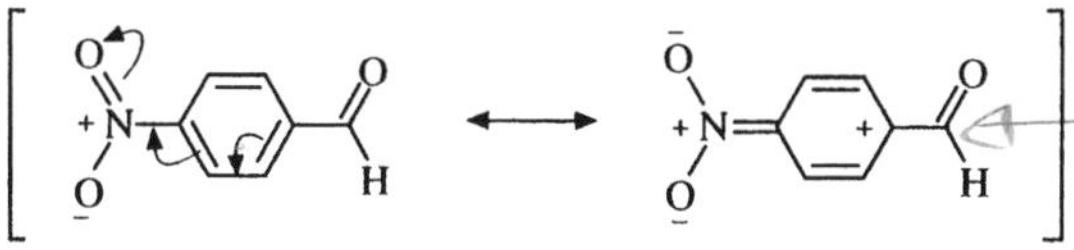

[handwritten: e⁻ deficient C atom ∴ very reactive towards nucleophiles.]

Aldehydes and ketones which are αβ-unsaturated are similarly deactivated towards nucleophilic attack at the carbonyl carbon, relative to the corresponding saturated compounds.

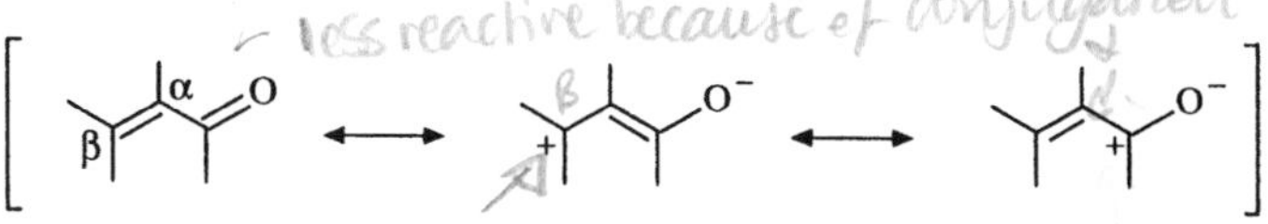

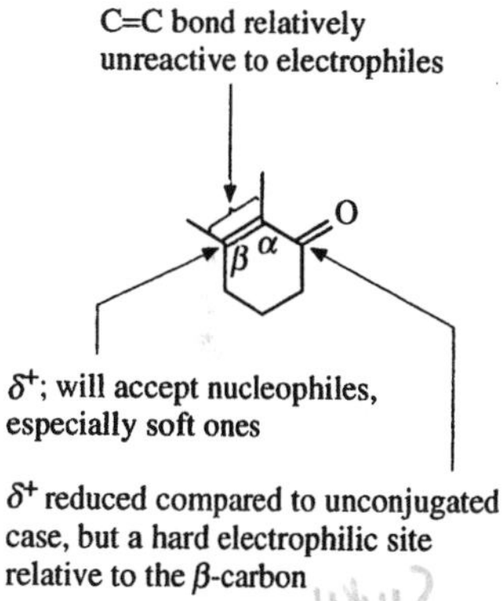

[handwritten: less reactive because of conjugation]

By the same token, the alkene unit here is less reactive versus electrophiles than otherwise similar but isolated carbon–carbon double bonds, and indeed acquires susceptibility to attack by nucleophiles at the β-carbon. Looked at another way, the effect of the electronic interaction between the two unsaturated groups is to share reactivity towards nucleophiles between them.

C=C bond relatively unreactive to electrophiles

δ^+; will accept nucleophiles, especially soft ones

δ^+ reduced compared to unconjugated case, but a hard electrophilic site relative to the β-carbon *[handwritten: why?]*

[handwritten: less steric hindrance?]

2.2 Hydration

Both aldehydes and ketones are reversibly hydrated in aqueous solution. At 20 °C in pure water, formaldehyde is almost 100 per cent hydrated at equilibrium; acetaldehyde is about 60 per cent hydrated; but acetone remains almost 100 per cent in the unhydrated form. The presence of the hydrated form in aqueous acetone is easily demonstrated, however, by the addition of $H_2^{18}O$, when ^{18}O is rapidly incorporated into the carbonyl group.

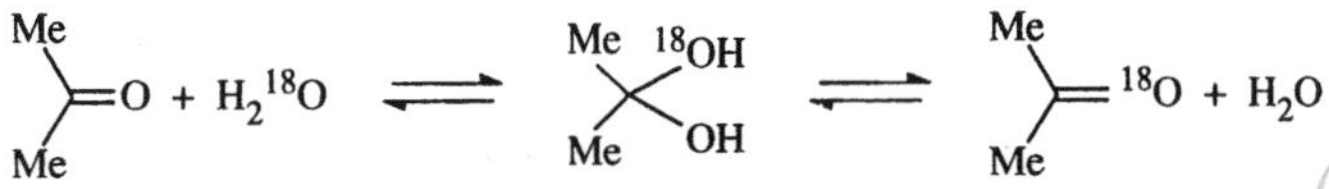

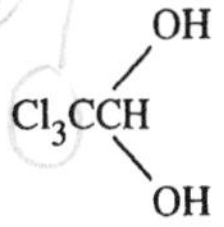

[handwritten: to make e⁻ more electrophilic]
[handwritten: e⁻ withdrawing group.]

Chloral hydrate

High degrees of hydration at equilibrium generally result when the carbonyl group has strong electron-withdrawing substituents attached, and in extreme cases it is the hydrate which is the familiar form of the carbonyl compound. The soporific chloral hydrate, and the α-amino-acid colorimetric reagent ninhydrin, for example, are both really fully hydrated forms of carbonyl compounds. In the case of ninhydrin, two intramolecular hydrogen bonds may contribute to the stability of the hydrate. Another example of this can be seen in 1,1-dihydroxycyclopropane: dehydration to cyclopropanone is difficult

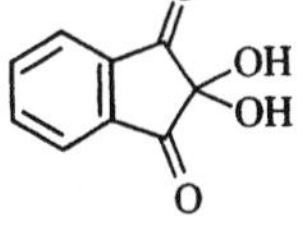

[handwritten: hydrogen bonding]

Ninhydrin

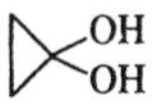

1,1-dihydroxycyclopropane

because it involves rehybridization from sp^3 to sp^2, and a corresponding hike in the angle strain. The ring bond angle of a tetrahedral carbon in a three-membered ring is already deformed by 40° from its relaxed value of 109°, but for a trigonal carbon (relaxed angle 120°) the deformation is 60°.

Carbonyl hydration is subject to catalysis by both acids and bases. The acid-catalysed pathway can be represented as follows.

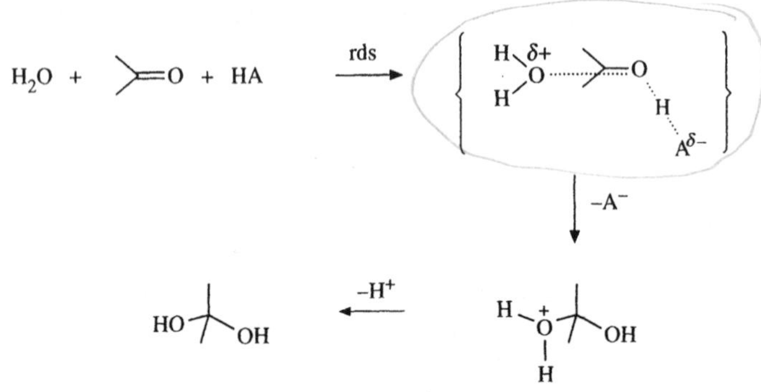

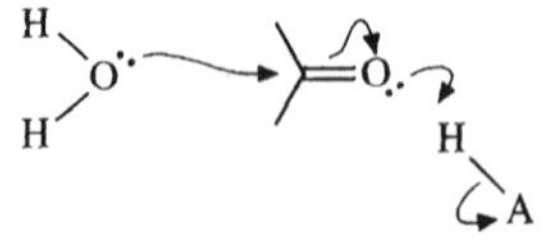

The transition state here entails the pull of electron density through the carbonyl oxygen, with concerted electron supply by the nucleophile to the carbonyl carbon. Because it is HA which is involved in the transition state, the rate of the reaction depends on [HA], or, if there are several catalysing acids, on Σ[HA]—not on [H$^+$]. This is called general acid catalysis: the key feature is that a proton transfer is part and parcel of the rate-determining step. The base-catalysed pathway can be sketched similarly.

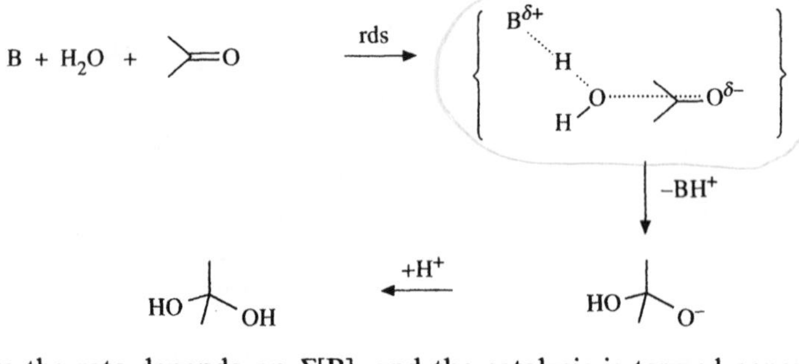

Here the rate depends on Σ[B], and the catalysis is termed general base catalysis. The reverse reactions are also subject to general acid and general base catalysis.

2.3 Hemiacetal and hemiketal formation

Hemi- is a prefix meaning half-. On the acetal ketal terminology see the marginal remark at the start of Chapter 3.

The 1:1 reaction of alcohols with aldehydes and ketones, giving hemiacetals and hemiketals, respectively, is closely analogous to their hydration.

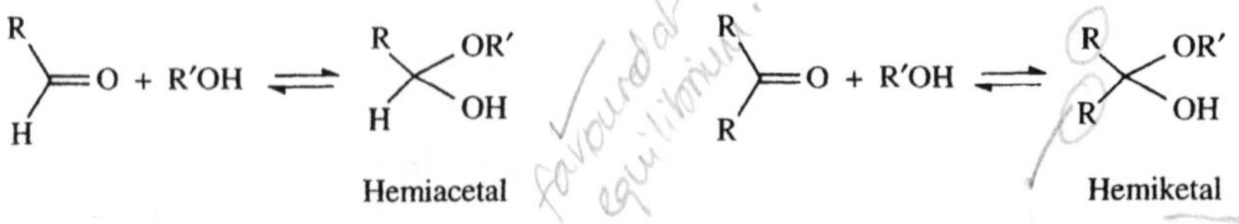

It is subject to general acid and general base catalysis. The adducts are favoured by electron-withdrawal from the carbonyl group, and, as with hydration, may be isolated in some cases, e.g. after treatment of polyhalo-aldehydes and -ketones with alcohols. But the equilibrium constants are somewhat less in favour of the adducts than for the analogous hydrations—about 0.5 for MeCHO/EtOH compared to about 1.5 for MeCHO/H_2O, for example.

Carbohydrates such as glucose provide important special instances of hemiacetal formation. D-Glucose has aldehydic and other chemical properties consistent with the extended acyclic formulation shown on the left in the next two schemes, but practically none of this structure is actually present in aqueous solution, because of spontaneous and practically complete cyclic hemiacetal formation, which gives two pyranose diastereoisomers.

Base catalysis ⊔ *Hemiacetal,*
acid catalysis ⊔ *acetal.*

acyclic aldehydic form

CHO
H—OH
HO—H
H—OH
HO—H
CH_2OH

β-D-Glucopyranose

α-D-Glucopyranose

equilibra to the RHS.

A pyran ring is one of 5 carbons and one oxygen.

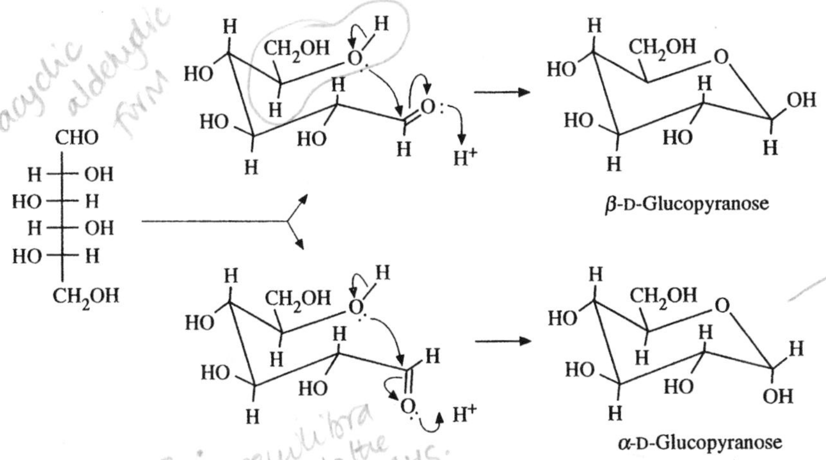

read A2 Biology.

read Dr Gabor's notes

This is a much more favourable process than intermolecular hemiacetal formation, because the reacting functionalities are tied together, and the six-membered pyranose ring is essentially strain-free. Intramolecular hemiacetal formation can also occur to give a five-membered furanose ring, again in two diastereoisomeric forms.

less steric hindrance + angle strain

Intramolecular

CHO
H—OH
HO—H
H—OH
HO—H
CH_2OH

β-D-Glucofuranose

α-D-Glucofuranose

A furan ring is one of 4 carbons and one oxygen.

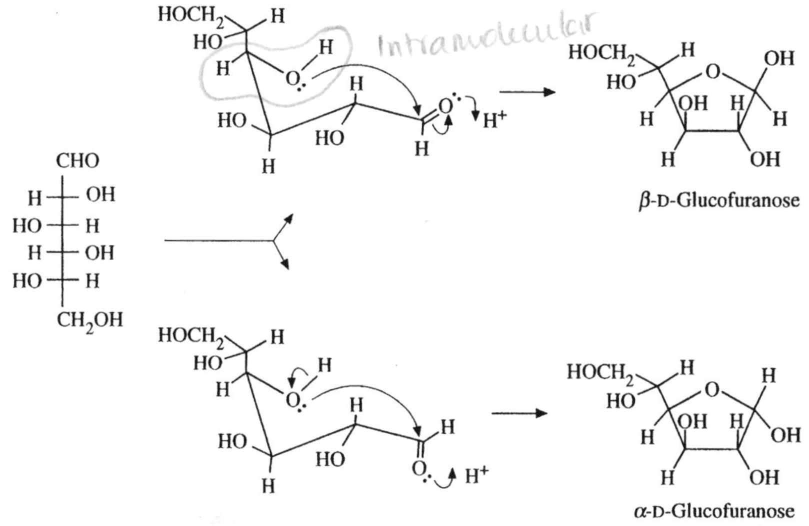

diastereoisomers)

An aqueous solution of D-glucose is thus a rather complex system, containing at equilibrium five solutes: the β- and α-pyranose forms (about 40 per cent and 60 per cent of the total, respectively), the β- and α-furanose forms (less than 1 per cent together) and the acyclic aldehydic form (a trace only). The ring-closure reactions are all reversible, freely and rapidly if acid base catalysis is available. If one of the solutes is removed from the solution, the equilibrium composition is rapidly restored. D-Glucose solutions in water thus give positive results with tests and reactions which are characteristic of aldehydes, despite the fact that the acyclic aldehydic form is only present in trace amounts.

2.4   Cyanohydrin formation

$$PhCHO + HCN \rightleftharpoons PhCH\overset{\displaystyle OH}{\underset{\displaystyle CN}{}}$$

Mandelonitrile

Aldehydes and ketones, except those which are sterically hindered or electronically deactivated, react with HCN, as in the marginal example. The reaction is base-catalysed and reversible; its mechanism is of historical interest, as it was the first reaction of organic chemistry to be interpreted (Lapworth, 1903) along modern lines.

Pure HCN reacts only very slowly, because it is a weak acid (pK_a about 9), but the reaction proceeds easily if base or a trace of CN^- is added. Because HCN is so hazardous—it is lethally poisonous and very volatile (b.p. 26 °C)—the reaction is usually performed by mixing the carbonyl substrate with NaCN and then adding sufficient acid to liberate the required amount of HCN *in situ*. The cyanide ion has electron density at both ends, either of which can react, according to the type of electrophile: with carbonyl groups it reacts through its more polarizable or 'soft' end. Note that in this reaction a new carbon–carbon bond is formed by causing a nucleophilic carbon atom to react with a complimentary electrophilic carbon atom. As such, it can be regarded as a prototype for a diverse and very important class of synthetic reactions which we shall have more to say about in Chapter 9. Here we will mention only one close analogy, the sodamide-induced reaction of acetylene with acetone in liquid ammonia. As the pK_a of ammonia (35) is greater than that of acetylene (25), the acetylene loses a proton to NH_2^-, giving an acetylide ion, which is isoelectronic with cyanide and reacts with electrophiles similarly.

Conditions: $HC\equiv CH/Me_2CO/NaNH_2$
in liquid ammonia; H_3O^+ work-up

2.5 Bisulfite addition

Sodium bisulfite reacts reversibly with aldehydes, less readily with methyl ketones, and not easily if at all with other ketones.

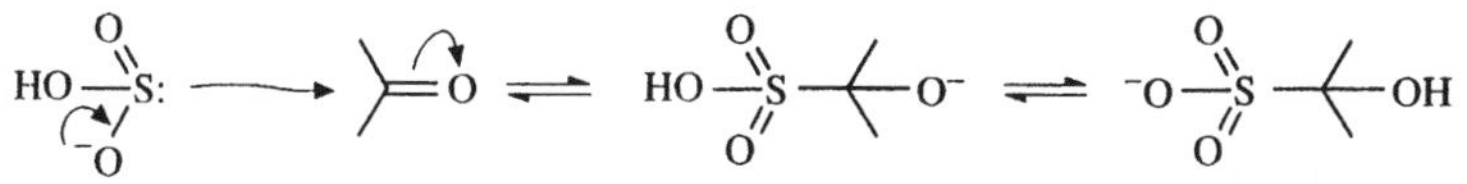

No catalyst is required, as the nucleophile is probably HSO_3^-, which is present already, reacting through its soft sulfur despite the fact that there are three potentially nucleophilic and apparently more accessible oxygens. It is a rather large nucleophile—hence its selectivity—and in general the equilibrium constants for the addition are less favourable to the adducts than is the case with cyanohydrin formation. The equilibria are shifted back towards starting materials by aqueous acid or base. Because the adducts are sulfonate salts, they are water-soluble, so aldehydes and reactive ketones may be extracted from organic solvents with aqueous sodium bisulfite and then regenerated by acidification or basification—a useful trick for their isolation.

2.6 Complex metal hydride reductions

Lithium aluminium hydride, 'lithal' in lab jargon, reacts readily and irreversibly with aldehydes and ketones in aprotic media. Aqueous work-up gives alcohols, formally the products of successive attack by H^- and H^+.

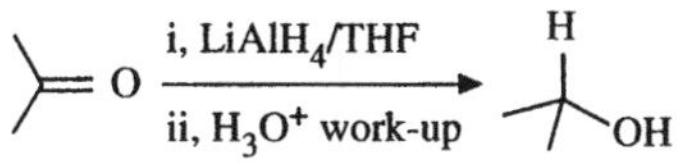

The AlH_4^- complex anion can be regarded just as a reservoir of nucleophilic H^-, but there is in reality much more to it than that. Sodium hydride—an ionic hydride, Na^+H^-—is not a nucleophilic reducing agent, but acts on aldehydes and ketones only as a very strong base. The distinctive feature of $LiAlH_4$ is that the source of nucleophilic H^- is accompanied by Lewis acids—the lithium cation and the aluminium atom—which are participants, not mere spectators. The key first step can be sketched as involving hydride transfer with both nucleophilic push and electrophilic pull. A somewhat speculative mechanism for the whole process is shown below, each successive hydride transfer being slower than the one before.

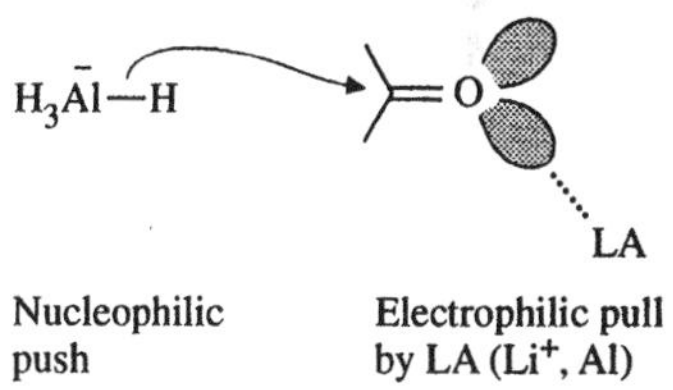

Nucleophilic push

Electrophilic pull by LA (Li^+, Al)

bulky ∴ sensitive to steric effect

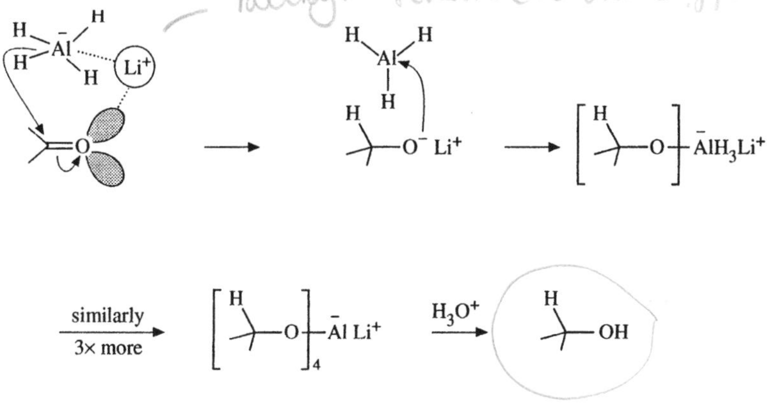

reactions with LiAlH₄ will undergo stereoselectivity, if required, due to its bulky nature

Because the nucleophilic entity is bulky, the reaction is sensitive to steric effects, and because it is irreversible there is stereoselectivity for hydride delivery to the least hindered side of an unsymmetrical cyclic ketone, even if that gives a compressed product.

stereoselectivity

Camphor   Borneol

Conditions: i, LiAlH$_4$/THF ii, H$_3$O$^+$ work-up

Least hindered side approach is also preferred with aldehyde and acyclic ketone substrates, although it is not so easy to see which side is which in such cases; see Section 2.10.

LiAlH$_4$ is a very reactive (and consequently dangerous) reagent which reduces many other functional groups containing electrophilic centres—carboxylic acid derivatives, epoxides, nitriles, and nitro compounds, for example.

The observation that the successive stages of aldehyde and ketone reduction by LiAlH$_4$ are each slower than the one before (see above) shows that lithium alkoxyaluminium hydrides are less reactive than LiAlH$_4$ itself. This can be turned to advantage, for such alkoxyaluminium hydrides are easily prepared by reaction of LiAlH$_4$ with controlled amounts of alcohols, and their moderated reactivity gives greater selectivity. Lithium tri-*t*-butoxyaluminium hydride, for example,

$$\text{LiAlH}_4 \ + \ 3\,\text{Bu}^t\text{OH} \longrightarrow \text{Li[AlH(OBu}^t)_3] \ + \ 3\,\text{H}_2\uparrow$$

is still reactive enough to reduce aldehydes and ketones at room temperature, but is very slow to attack ester, nitrile and nitro groups. LiAlH$_4$ is inert towards isolated carbon–carbon double bonds, but such bonds are attacked if they are conjugated with a carbonyl group, and $\alpha\beta$-unsaturated ketones are attacked at both electrophilic sites (see Section 2.1) to give mixtures.

LiAlH$_4$ attack only if conjugated with carbonyl

85%   15%

Conditions: LiAlH$_4$/Et$_2$O/–10 °C; H$_3$O$^+$ work-up

Sodium borohydride reduces aldehydes and ketones by close analogy with LiAlH$_4$, but more smoothly and selectively. It is the reducing agent of choice except for unreactive substrates. Whereas LiAlH$_4$ reacts violently with water, and cannot be used except in aprotic conditions, NaBH$_4$ is sufficiently kinetically stable for it to be employed in methanol etc., and even in water itself. Most esters, nitriles etc., are attacked very slowly or not at all, but, as with LiAlH$_4$, $\alpha\beta$-unsaturated ketones may be reduced at the carbon–carbon double bond as well as the carbonyl group. Thus excess $NaBH_4/EtOH$ reduces cyclopentenone completely to the saturated alcohol. The carbon–carbon double bond must be reduced first in this case, as the allylic alcohol which would otherwise be an intermediate is not affected under the same conditions. The selectivity can be inverted, however, by the addition of cerium trichloride, when only the carbonyl group is reduced (the Luche reduction).

Conditions: i, $NaBH_4$(xs)/EtOH; ii, $NaBH_4$–$CeCl_3$/MeOH

Too far away for internal H^- delivery

The high selectivity is probably due to internal delivery of the reducing hydride in a tightly bound ceric ion complex.

When aldehydes are treated with $NaBH_4$ in aqueous or alcoholic media, the formation of hydrates and/or hemiacetals is of no practical consequence, because the addition of water and alcohols is freely reversible. Only the free aldehyde is reduced, but its consumption causes more to be released by displacement of the equilibrium, until all the starting material has reacted. Aldehydes have greater intrinsic reactivity towards nucleophiles than ketones. They are consequently reduced more easily by $NaBH_4$, despite the much easier addition of water and alcohols to the former. Inverse selectivity can nevertheless be achieved, again by the clever use of Ce^{3+}, which probably coordinates to and stabilizes aldehyde hydrates, thereby preventing reduction, but leaves essentially unhydrated ketones exposed.

Conditions: i, $CeCl_3$/aq.EtOH; ii, xs $NaBH_4$; iii, xs Me_2CO; iv, aq.NaCl

2.7 The Meerwein–Ponndorf–Verley reaction

If a relatively involatile ketone such as cyclohexanone is heated with aluminium isopropoxide in isopropanol, the corresponding alcohol is obtained, after work-up with aqueous acid. This is the Meerwein–Ponndorf–Verley reaction.

Reversible hydride transfer via a comfortable six-membered transition state is involved here, equilibrium being displaced to the right by distilling out the acetone which is formed.

Distilled out, displacing
equilibrium to the right

The process can be driven in the opposite direction too, when it is called the Oppenauer oxidation: treatment of a secondary alcohol with $Al(OBu^t)_3$ and excess acetone gives the corresponding ketone, after aqueous acid work-up.

$$R_2CHOH \xrightarrow[\text{ii, } H_3O^+ \text{ work-up}]{\substack{\text{i, } \Delta/Me_2CO(xs)/\\ Al(OBu^t)_3}} R_2C=O$$

2.8   The Cannizzaro reaction

In a reaction discovered by Cannizzaro nearly one hundred and fifty years ago, benzaldehyde undergoes formal disproportionation on treatment with strong alkali.

$$2PhCHO \xrightarrow[\substack{\text{ii, } H_3O^+\\ \text{work-up}}]{\text{i, } \Delta/NaOH} \substack{PhCO_2H\\ +\\ PhCO_2H}$$

This is a hydride-transfer reaction, in which the hydride source is generated by reversible nucleophilic addition of OH^- to the carbonyl group, the resultant tetrahedral adduct undergoing rate-determining reaction with another benzaldehyde molecule.

For such a scheme, it would be expected that the rate of the overall reaction would be proportional to [adduct][PhCHO]; but since the adduct is formed by a freely reversible equilibrium from PhCHO and OH^-, its concentration will always be proportional to [PhCHO][OH^-], so the overall reaction rate should be proportional to $[PhCHO]^2[OH]^-$. This is in fact what is observed experimentally. The Cannizzaro reaction is not a general reaction of aldehydes, however, because if α-CH is present other pathways are preferred (see Chapter 9). Furthermore, crossed reactions only work in one direction in special circumstances. One such case is the reaction of formaldehyde with benzaldehyde in the presence of alkali.

$$H_2CO + PhCHO \xrightarrow[\text{ii, } H_3O^+ \text{ work-up}]{\text{i, } \Delta/\text{NaOH}} PhCH_2OH + HCO_2H$$

Here the formaldehyde, which has by far the more reactive carbonyl group, is quantitatively converted to the hydride-donating adduct, which can therefore only react with an unchanged benzaldehyde molecule, of which an abundance is available, since its equilibrium reaction with OH⁻ is much less favourable to the adduct.

2.9 Reaction with Grignard reagents

Alkyl and aryl chlorides, bromides and iodides react with pure, clean magnesium turnings in pure, dry diethyl ether or similar ethereal solvents like THF. The resulting solutions, which are called Grignard reagents (or often just 'Grignards') after their discoverer, are complex systems, but the stoichiometry of the reaction is simple, and the naive formulations R–MgHal or R⁻[MgHal]⁺ suffice to rationalize the reactions with which we are concerned. Grignards react with aldehydes and ketones to give primary or secondary alcohols, respectively, after aqueous acid work-up.

$$R-Hal + Mg \longrightarrow RMgHal$$

Conditions: pure dry reagents in pure dry Et₂O or THF

R—MgBr

$$RMgHal + \; \rangle{=}O \longrightarrow R \longrightarrow OMgHal \xrightarrow{H_3O^+} R \longrightarrow OH$$

The reactions involve both electron supply by the nucleophile (nominally R⁻), and electrophilic pull by [MgHal]⁺, the Lewis acid component of the reagent.

This process bears close comparison with the reduction of aldehydes and ketones by complex metal hydrides, and like that reaction is sensitive to steric hindrance. With some ketones, side-reactions provide easier pathways than nucleophilic attack at an obstructed carbonyl carbon. If α-CH is present, it is likely to be more accessible, and the reagent may then act as a base instead of as a nucleophile, leading to enolate formation and regeneration of the starting ketone on aqueous acid work-up.

Nu = attack the C atom directly.

base = attack the XH atom next to C.

(ii) contiguous H atom attack leads to enolate.

Starting material recovered on work-up

Alternatively, if a β-CH is available in the reagent, reduction by hydride-transfer cf. the Meerwein–Ponndorf–Verley reaction) to the carbonyl carbon may take place instead of attack by the nucleophilic carbon.

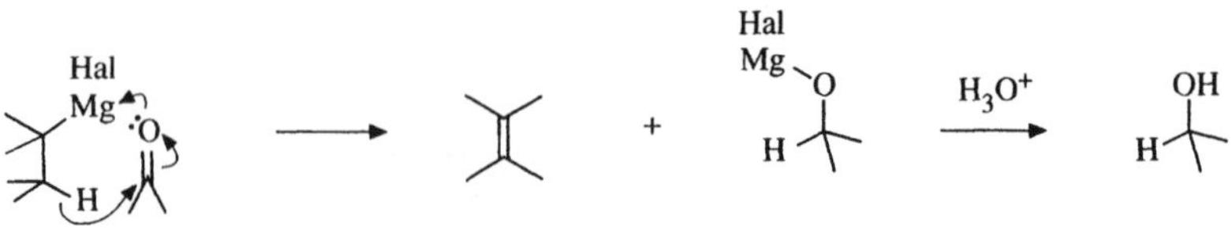

Thus isopropylmagnesium bromide reacts with diisopropyl ketone to give propane, propene, diisopropylmethanol, and recovered ketone after aqueous work-up, but no triisopropylmethanol.

$$Pr^i_3C\text{–}OH \;\not\longleftarrow\; Pr^iMgBr + Pr^i_2C{=}O \xrightarrow[\text{work-up}]{\text{after}} \left\{ \begin{array}{l} Pr^i_2CHOH \\ + \\ CH_3CH{=}CH_2 \\ + \\ CH_3CH_2CH_3 \\ + \\ Pr^i_2C{=}O \quad \text{recovered} \end{array} \right.$$

Organolithium reagents are less easily diverted into side reactions by steric hindrance than Grignards are: isopropyllithium reacts with diisopropyl ketone to give triisopropylmethanol in good yield.

$$Pr^iLi + Pr^i_2C{=}O \xrightarrow[\text{work-up}]{\text{after}} Pr^i_3C\text{–}OH$$

Organocerium reagents, prepared *in situ* from Grignards and ceric chloride, are also less prone than Grignards on their own are to yield enolization and reduction products on reaction with ketones.

$$RMgBr + CeCl_3 \longrightarrow RCeCl_2 + MgBrCl$$

Methyl mesityl ketone, for example, gives a derisory yield of tertiary alcohol on treatment with *n*-butylmagnesium bromide alone, but the yield is satisfactory if ceric chloride is added.

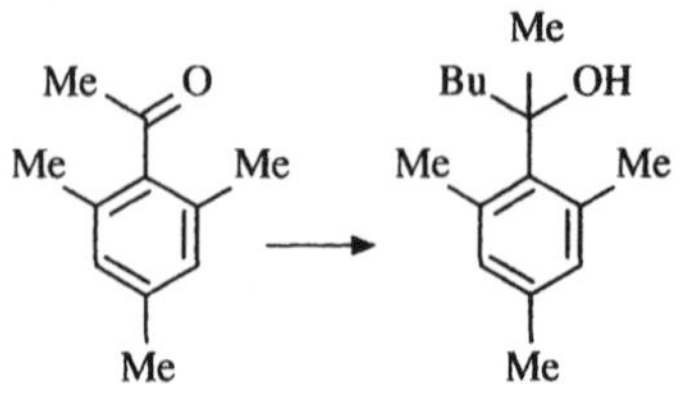

Conditions: $BuCeCl_2$/THF/–78 °C

Grignards react at both electrophilic sites in $\alpha\beta$-unsaturated ketones; alkyllithium reagents generally attack the carbonyl group only; lithium dialkylcuprates, on the other hand, attack only at the β-position.

Conditions: i, MeMgBr; ii, MeLi; iii, Me$_2$CuLi; H$_3$O$^+$ work-ups

This selectivity above can be interpreted in hard and soft terms. The relatively ionic methyllithium reagent is practically a source of a hard Me$^-$ nucleophile, which seeks out the harder of the electrophilic sites on the substrate; the more covalent MeMgBr presents a methyl nucleophile which is of intermediate and undiscriminating hardness/softness; but the large and polarizable copper(I) atom makes the Me$_2$Cu$^-$ nucleophilic moiety in Me$_2$CuLi soft and selective for the softer electrophilic substrate site. This picture, especially in relation to the copper-containing reagent, which exhibits structural and mechanistic complexity, is undoubtedly too simplistic, but it does provide a convenient rationalization of the facts.

2.10 Stereochemical aspects of nucleophilic addition to aldehydes and acyclic ketones

When an achiral nucleophilic reagent adds to an achiral aldehyde or simple acyclic unsymmetrical ketone, a chiral centre is formed, but the two enantiomers are of course formed in equal amount. Thus the reaction of phenylmagnesium bromide with methyl ethyl ketone gives a racemic tertiary alcohol, methyl ethyl phenyl carbinol. However, if a chiral centre already exists near the carbonyl group, there are two possible diastereoisomeric products, and these are in general not formed in equal amounts.

With freely reversible nucleophilic additions like cyanohydrin formation, thermodynamic differences between the diastereoisomeric products—which may be slight in simple cases—will determine which will predominate. With irreversible processes such as LiAlH$_4$ reduction or reaction with Grignard reagents, on the other hand, kinetic considerations depending mainly on steric factors will determine the outcome.

Consider a carbonyl compound, where L is the largest group, M is a group of intermediate size, and S the smallest group. If a Newman projection is made along the bond between the chiral α-carbon and the carbonyl carbon, and the carbon–oxygen double bond is placed anti to

Nucleophilic attack least hindered on this side

the substituent L, then Cram's rule predicts the stereoselectivity by supposing that this is the reactive conformation, and that the nucleophile will attack the electrophilic carbon most easily on the side where the smallest α-substituent S is. Thus (R)-3-phenylbutan-2-one gives about 75 per cent of the (R,S)-3-phenyl-2-hydroxybutane and 25 per cent of the (R,R)-diastereoisomer on treatment with LiAlH$_4$, followed by aqueous work-up as usual.

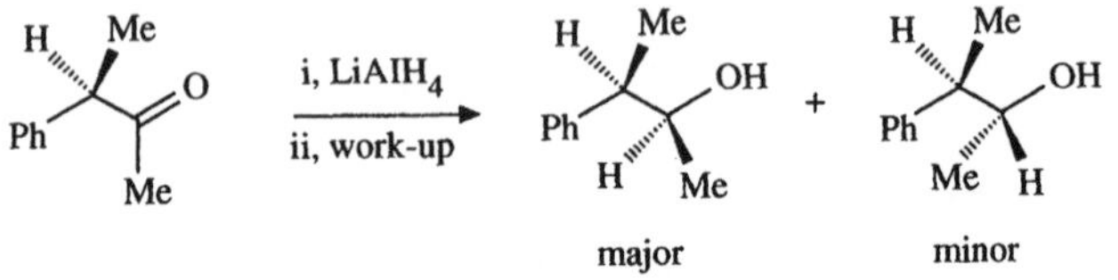

major   minor

Cram's rule is a useful general guide, despite the fact that it is based on a model for the reactive conformation which is no longer favoured, but it breaks down if nonsteric factors enter into consideration. For example, if there is an α-substituent which can join forces with the carbonyl oxygen to chelate a metal cation, the conformation of the substrate is constrained. Nucleophilic attack still takes place on the least hindered side, but this may or may not lead to the product predicted by Cram's rule.

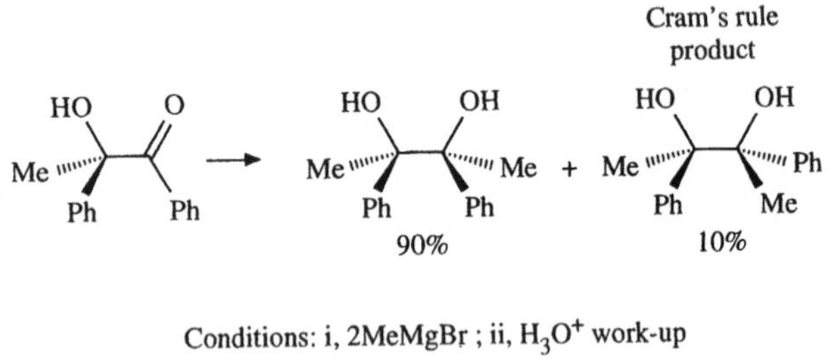

Cram's rule product

90%   10%

Conditions: i, 2MeMgBr ; ii, H$_3$O$^+$ work-up

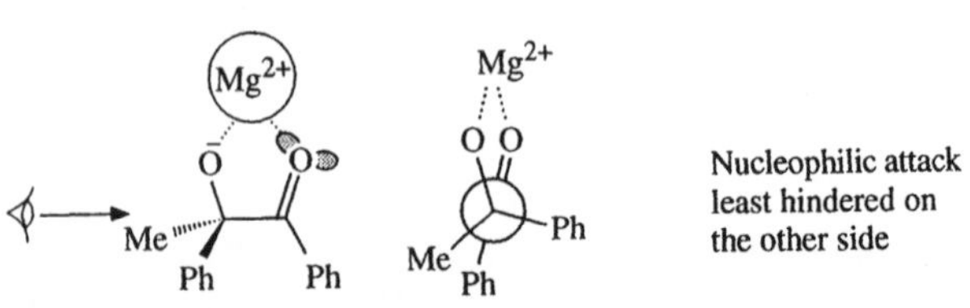

Nucleophilic attack least hindered on the other side

The rationale behind Cram's rule as first postulated was that the carbonyl oxygen, being in effect enlarged by coordination to Lewis acids, would seek to be as far from the group L as possible. It is now thought more plausible—this is the Felkin–Anh model—to suppose that lowest energy transition state (i.e. the one leading to the favoured diastereoisomer) will be derived from the substrate conformation in which the group L distances itself equally from the R group and the carbonyl oxygen, with the group S lying close to the trajectory of the nucleophile. The predicted outcome is, however, the same as that arrived at with Cram's original model.

> 90° to the carbonyl group.

Nucleophilic attack least hindered on this other side

Cram's rule: largest group antiperiplanar to the carbonyl group.

Problems

Explain the following.

1.

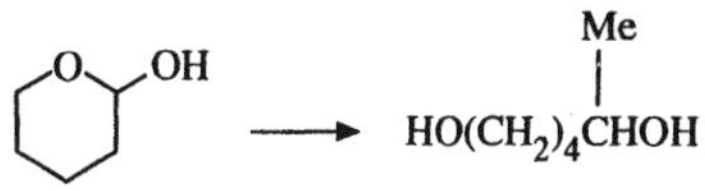

$$HO(CH_2)_4\overset{\overset{\displaystyle Me}{\displaystyle |}}{C}HOH$$

Conditions: i, xs MeMgBr; ii, H_3O^+ work-up

2.

Conditions: i, HCN; ii, MeOH/H_2SO_4

3.

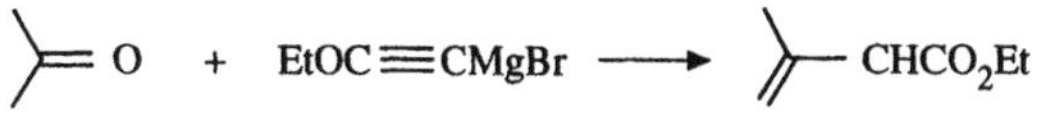

+ EtOC≡CMgBr ⟶

Conditions: i, reaction in Et_2O; ii, H_3O^+ work-up

3   Acetals and ketals

3.1   Formation

We have already seen that aldehydes and ketones react reversibly 1:1 with alcohols under general acid or general base catalysis, to give hemiacetals and hemiketals, respectively (see Section 2.3). With excess alcohol and a catalytic amount of a strong acid, further reversible reaction takes place to replace the OH group and give an acetal (from an aldehyde) or ketal (from a ketone). The overall equilibria are as follows.

$$\underset{H}{\overset{R}{\diagup}}C{=}O \; + \; R'OH(xs) \; \rightleftharpoons \; \underset{H}{\overset{R}{\diagup}}C\underset{OR'}{\overset{OR'}{\diagdown}} \; + \; H_2O$$

Acetal

$$\underset{R}{\overset{R}{\diagup}}C{=}O \; + \; R'OH(xs) \; \rightleftharpoons \; \underset{R}{\overset{R}{\diagup}}C\underset{OR'}{\overset{OR'}{\diagdown}} \; + \; H_2O$$

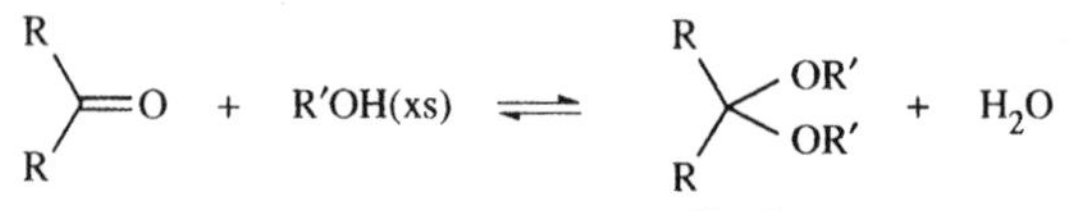

Ketal

If arrangements are made to remove the water from the continuously equilibrating mixture, by distillation or other means, then the aldehyde or ketone is quantitatively converted to an acetal or ketal, as the case may be. The reaction proceeds more easily with aldehydes than with ketones, and, in the simple form sketched above, is practically limited to primary alcohols. 1,2-Diols and 1,3-diols react easily, giving five- and six-membered rings, respectively.

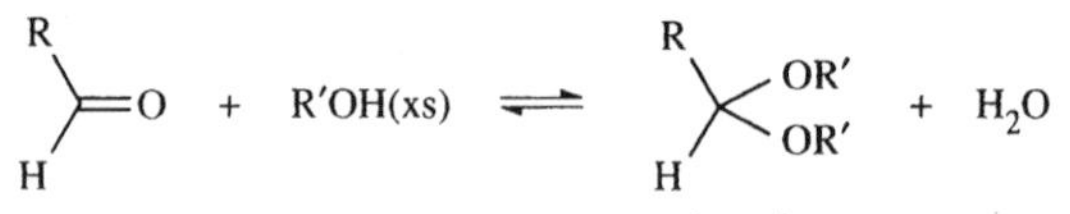

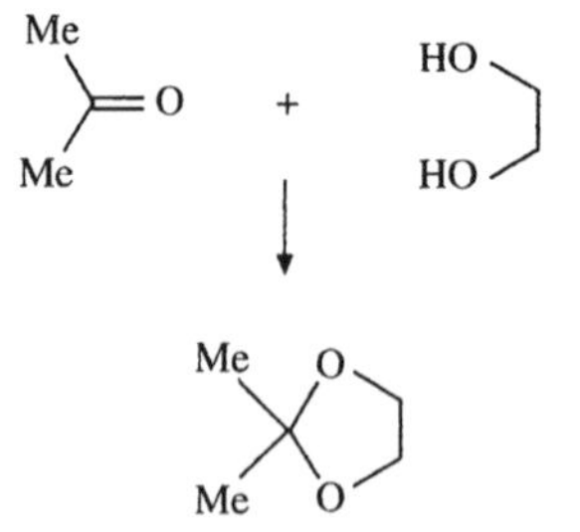

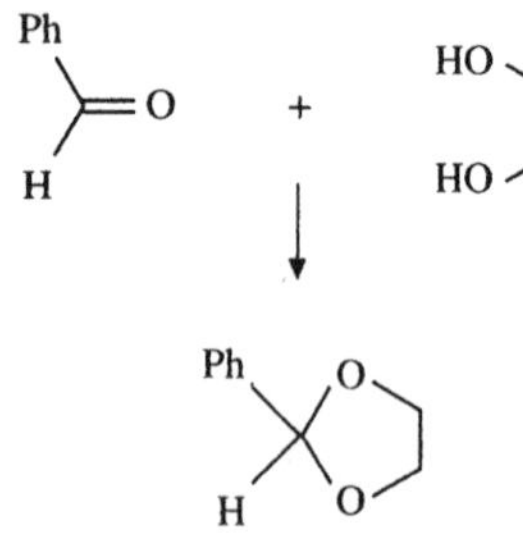

In the reactions of diols the favourability of ring-formation offsets the intrinsic unreactivity of secondary hydroxyl groups, and cyclic ketal and acetal formation is of great importance in carbohydrate chemistry. Thus D-glucose reacts with acetone in the presence of sulfuric acid to give diacetone glucose.

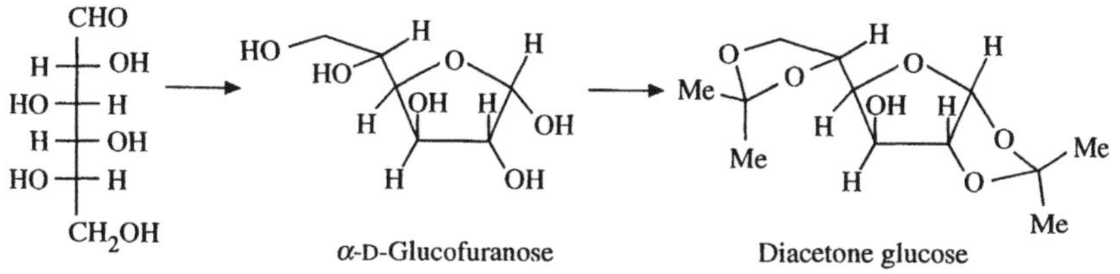

α-D-Glucofuranose Diacetone glucose

Conditions: Me$_2$CO/H$_2$SO$_4$/Δ

The α-furanose derivative shown is produced because thermodynamic control operates in the freely equilibrating reaction mixture. Only the α-furanose form of D-glucose has two *cis*-1,2-diol systems able to give the five-membered cyclic ketals which are especially favoured in reactions of acetone.

The acid-catalysed equilibration of D-glucose with simple alcohols gives a pair of epimeric cyclic mixed acetals.

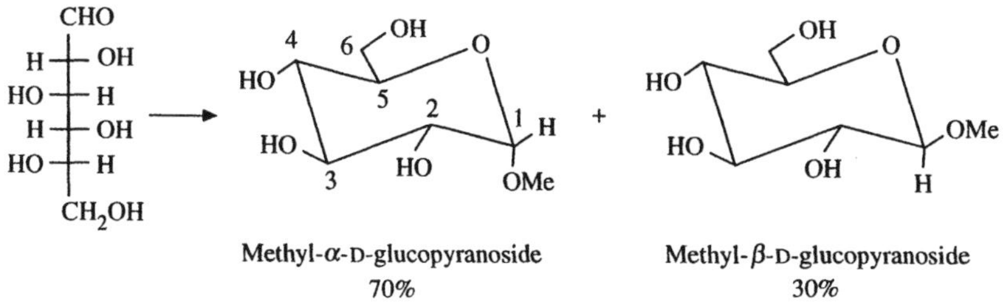

Methyl-α-D-glucopyranoside Methyl-β-D-glucopyranoside
70% 30%

Conditions: MeOH/H$^+$ to equilibrium

Note that although methanol is a primary alcohol, and present here in vast excess, it cannot compete with cyclization through the more hindered secondary OH group at C-5. Mixed acetal derivatives of carbohydrates involving C-1 are called glycosides. Many complex glycosides occur in living things. Cellulose, for example—the main constituent of plant cell walls—is a polymer in which β-D-glucopyranose units are connected by glycoside formation between C-1 of one unit and the OH at C-4 of the next. Here the mixed acetal construct serves both to maintain the rings and to join the units together.

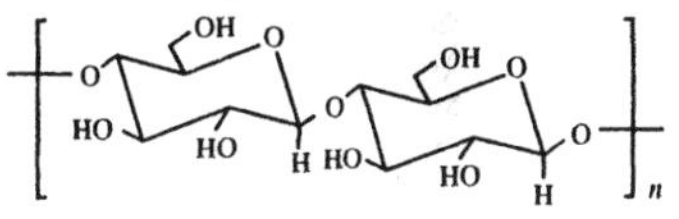

Cellulose

Acetal formation takes place by a multistep mechanism which follows on after hemiacetal formation as described in Section 2.3.

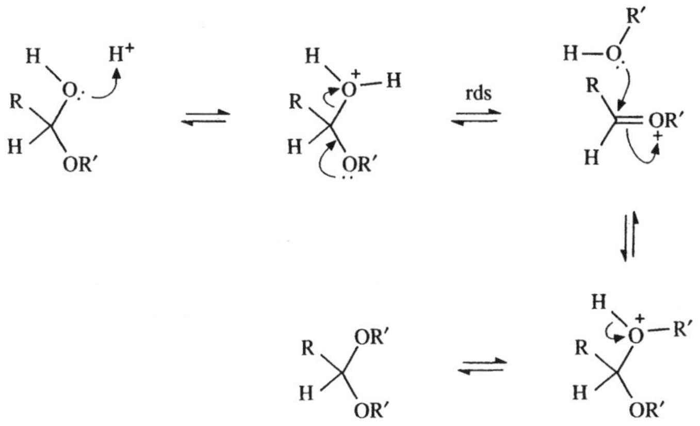

The rate-determining step is the unimolecular dissociation of the protonated hemiacetal. The rate of this step, and therefore of the overall reaction, depends only on [hemiacetal.H^+], which must depend in turn on [H_3O^+], i.e. on pH. Such catalysis, where the acidity of the medium is what matters, is termed specific acid catalysis, in contrast to general acid catalysis, where the total concentration of catalysing acids is the controlling factor. The key difference between the two types of catalysis is that in general acid catalysis a proton transfer from a catalysing acid to the substrate is taking place in the rate-determining transition state, but in specific acid catalysis the proton transfer which induces reaction is complete before the rate-determining stage is reached. The two types of acid catalysis can be distinguished experimentally by investigating the response of the reaction to buffer concentration at constant pH: ideally, a reaction subject to specific acid catalysis should show invariant rate, whereas one subject to general acid catalysis should accelerate with increasing buffer concentration. Ketal formation from ketones via hemiketals takes place by exactly the same mechanism as acetal formation from aldehydes via hemiacetals.

The polymerization which takes place on evaporation of an aqueous solution of formaldehyde (formalin), giving paraformaldehyde, is a special case of acetal formation.

Trioxane

Paraformaldehyde

Paraldehyde

Treatment of formaldehyde with sulfuric acid, on the other hand, gives a trimer, trioxane; acetaldehyde also gives a cyclic trimer (paraldehyde) or a cyclic tetramer (metaldehyde) on treatment with catalytic amounts of sulfuric acid, depending on the conditions. These acetals are all solids, which can be used as convenient sources of the aldehydes from which they are derived, to which they are readily hydrolysed with dilute aqueous acid.

Metaldehyde

3.2   Properties

The CH between the oxygen atoms in an acetal or ketal is of low acidity, because there is no scope for delocalization of the anion which would be formed. Ketals are completely stable to bases. So are acetals, to all except extremely basic reagents such as butyllithium, with which the acetal proton may be abstracted, as in the example below.

Since acetals and ketals are saturated compounds, nucleophilic attack can only occur if a carbon–oxygen or carbon–carbon bond is broken. This would involve displacement of alkoxide or alkyl anions, which are, respectively, poor and exceedingly poor leaving groups in the absence of special stabilizing structural features. Acetals and ketals are thus stable to most nucleophilic reagents, although carbon–oxygen bond fission can be induced by Grignard reagents with the assistance of added Lewis acids. This reaction takes place with difficulty, however, compared to other reactions of Grignard reagents, such as addition to carbonyl groups.

In complete contrast to the high stability of acetals and ketals to base, aqueous acid easily hydrolyses them by displacing the formation equilibria (see above) back to the carbonyl compounds from which they were derived.

3.3 Dithioacetals and dithioketals

See also Primer 33.

Thiols react with aldehydes and ketones under acidic conditions by analogy with alcohols, to give eventually dithioacetals or dithioketals, but the equilibria are more favourable to the adducts.

BF$_3$·Et$_2$O may also be used as catalyst for dithioketalization

Dithioacetals and dithioketals are not only stable to aqueous base like their oxygen counterparts, but also to aqueous acid: for hydrolysis to the corresponding carbonyl compounds, addition of mercury(II) compounds is required.

The CH between the heteroatoms in dithioacetals is rather more acidic than it is in acetals. Five-membered cyclic dithioacetals undergo the same fragmentation on treatment with alkyllithium reagents as five-membered cyclic acetals do (see above). A six-membered cyclic dithioacetal is not set up to fragment in this way, however, and gives a stabilized organometallic intermediate, which is highly nucleophilic at the central carbon atom. Treatment of this intermediate with an electrophile such as an alkyl iodide gives a dithioketal, from which a ketone can be obtained by mercury(II)-mediated hydrolysis. These reactions can be used to convert an aldehyde to a ketone.

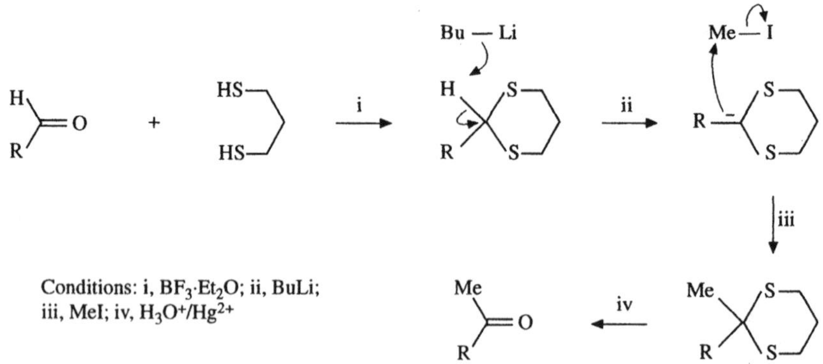

Conditions: i, $BF_3 \cdot Et_2O$; ii, BuLi; iii, MeI; iv, H_3O^+/Hg^{2+}

In this sequence, the electrophilic carbonyl carbon of the original aldehyde is temporarily made nucleophilic for the purposes of the synthesis, emerging finally electrophilic again in the ketone product. This relatively novel concept was developed in German-speaking parts, and there is a convenient German word for the inversion of polarity which is involved: *umpolung*, which has now become part of the English vocabulary of organic chemistry.

3.4   Orthoesters

Orthoesters, $RC(OR')_3$, of which the best known is ethyl orthoformate (R = H), are formally acetals derived from esters. They cannot be prepared from esters by analogy with aldehyde and ketone chemistry, because the intervening equilibria are unfavourable, but they can be accessed from other starting points, as in the following preparation of ethyl orthoformate.

$$NaOEt + CHCl_3 \longrightarrow NaCl + HC(OEt)_3$$

Conditions: Na metal added to $EtOH/CHCl_3$

Orthoesters are base-stable, but *O*-protonation or interaction with a Lewis acid breaks one of the carbon–oxygen bonds rather easily, because the resulting cation is resonance-stabilized.

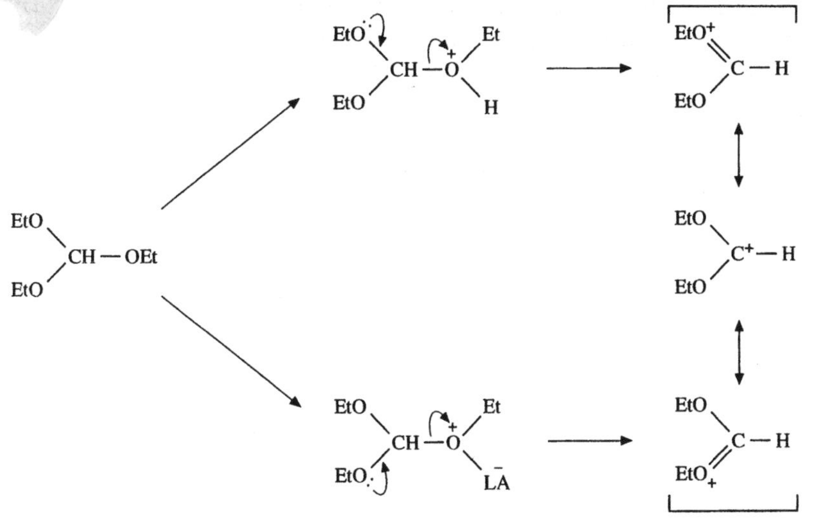

Hydrolysis via this cation under acidic conditions is very facile, giving first an ester, which undergoes further hydrolysis (see Chapter 5) under more vigorous or extended treatment.

$$HC(OEt)_3 \longrightarrow HCO_2Et + EtOH$$
Conditions: cold, dilute H_3O^+

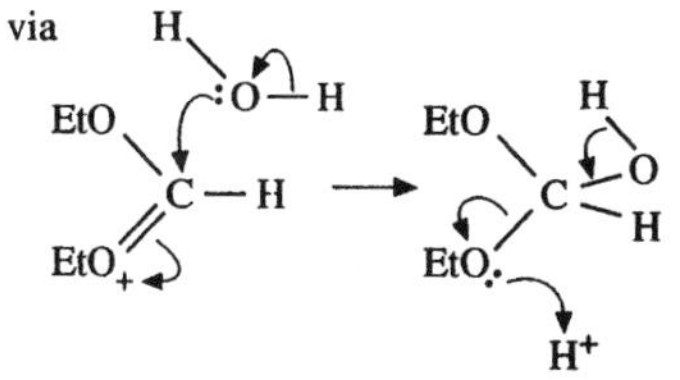

Similarly, Grignard reagents react with ethyl orthoformate to give acetals, which can, if desired, be hydrolysed as part of the work-up, yielding aldehydes.

$$RMgBr + HC(OEt)_3 \xrightarrow{\text{i}} RCH(OEt)_2 \xrightarrow{\text{ii}} RCHO$$

Conditions: i, reflux, Et_2O, several hours, then aq. AcOH work-up; ii, dilute aq. H_2SO_4/Δ

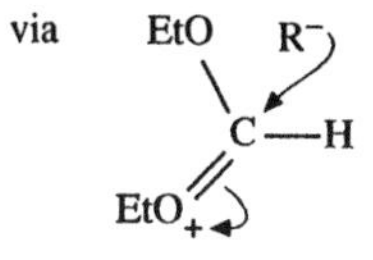

3.5   The protecting group principle

In elementary textbooks, the reactions of organic chemistry are generally illustrated for simplicity's sake with transformations of monofunctional compounds. In real chemical life, however, polyfunctionality is the norm, and a common problem is the selective execution of an interconversion at one functional centre when a second is present which might interfere, or itself undergo irreversible change under the intended conditions. In this situation, it is desirable to derivatize the second functionality in such a way that its reactivity is blocked whilst enabling it to be liberated subsequently on demand. A group employed for such a purpose is called a protecting group. Acetals and ketals are ideal for protecting aldehydes and ketones, because they can be prepared from them easily; they remain aloof from all except extremes of nucleophilic and basic reagents; and they yield up the original carbonyl groups when required on mild acidic hydrolysis. For example, suppose that it is desired to reduce an ethyl ester group with LiAlH$_4$ (here we anticipate a little: see Section 5.5) without affecting a methyl ketone group elsewhere in the molecule. As posed, this is an impossible task, because, as we have seen (Section 2.6), LiAlH$_4$ also reduces ketones readily. Protection of the ketone as a ketal provides the solution, and enables the desired ester reduction to be achieved with complete selectivity, albeit by slightly roundabout means.

For examples of group protection from the peptide field see Primer 7.

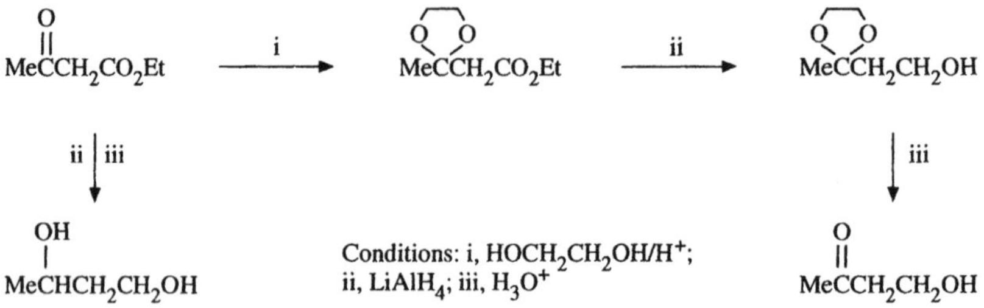

Conditions: i, HOCH$_2$CH$_2$OH/H$^+$;
ii, LiAlH$_4$; iii, H$_3$O$^+$

Problems

Explain the following.

1.

Conditions: i, BuLi; ii, Cl(CH$_2$)$_3$Br; iii, BuLi; iv, H$_3$O$^+$/Hg^{++}

2.

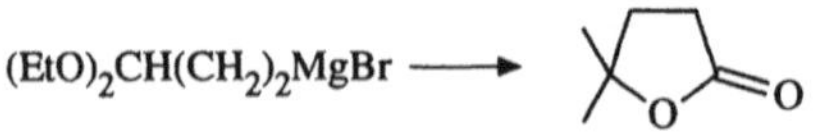

(EtO)$_2$CH(CH$_2$)$_2$MgBr ⟶

Conditions: i, Me$_2$CO; ii, H$_3$O$^+$; iii, H$_2$CrO$_4$

4 Reactions of amino compounds with aldehydes and ketones

4.1 Introduction

Amino nucleophiles $\sim NH_2$ react reversibly with aldehydes and ketones.

See also Primer 38.

Nitrogen stands beside oxygen in the periodic table, so NH_3 and $\sim NH_2$ can be regarded as structural analogues of water and alcohols, respectively. The reversible reaction of $\sim NH_2$ with an aldehyde or ketone is very similar to that of water, involving as it does nucleophilic addition followed by dehydration. A general mechanism can be outlined as shown below.

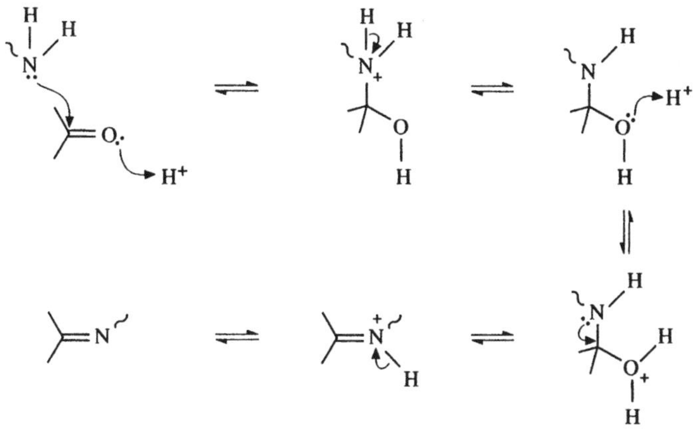

Secondary amines R_2NH are intrinsically just as nucleophilic as primary amines RNH_2. They react with aldehydes and ketones similarly, except that the final stage is blocked because there is no proton to discard, so iminium ions—intermediates in enamine formation (see Section 4.4) and in the Mannich reaction (see Section 9.3)—are generated.

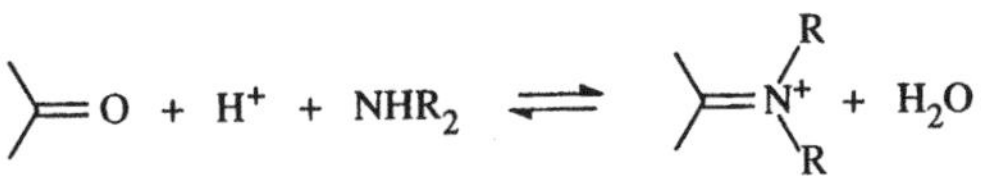

4.2 Classical aldehyde and ketone derivatization

Perhaps the best-known reactions of the above type are those of 2,4-dinitrophenylhydrazine, semicarbazide, and hydroxylamine with aldehydes and ketones, which give 2,4-dinitrophenylhydrazones, semicarbazones, and oximes, respectively. These derivatives (Fig. 4.1) which

2,4-Dinitrophenylhydrazone

Semicarbazone

Oxime

can be made by simple test-tube procedures, are in most cases nicely crystalline compounds with well-defined and characteristic melting points. They have been valued for the identification of aldehydes and ketones since the early days of organic chemistry. The case of acetoxime formation has been thoroughly studied, and is instructive. The rate of the reaction depends critically on the pH. A graph of rate versus pH has a bell-shaped profile characteristic of many reactions in this class.

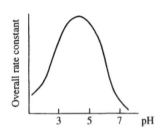

Acetoxime formation from acetone and excess hydroxylamine.

$$H^+ + \overset{\cdot\cdot}{N}H_2OH \rightleftharpoons \overset{+}{N}H_3OH$$

Nucleophilic   Non-nucleophilic

Acetoxime

In order to function as a nitrogen nucleophile, and to attack generating the tetrahedral intermediate, the hydroxylamine reagent must be present in the free base form, the concentration of which is lower the lower the pH. At low pH, the formation of the tetrahedral adduct is rate-determining. On the other hand, the dehydration of the tetrahedral intermediate to give the oxime product is acid-catalysed, and is therefore slow and rate-determining at high pH.

The two opposing trends reach an optimum compromise at intermediate pH, when there is both enough neutral nucleophile for the first stage to generate a supply of the tetrahedral intermediate and a sufficient concentration of protons for effective catalysis of its dehydration in the second stage.

N-Monosubstituted hydroxylamines also react as nitrogen nucleophiles with aldehydes and ketones, but in this case the tetrahedral adducts dehydrate to give nitrones (azomethine N-oxides).

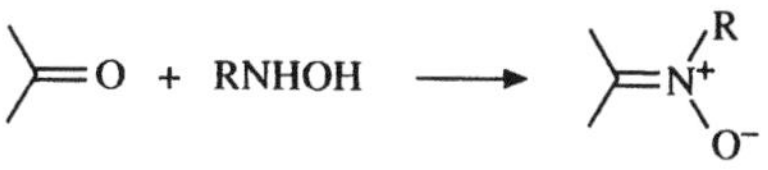

A nitrone

4.3 Reactions of ammonia and primary amines with aldehydes and ketones

Ammonia itself and primary aliphatic amines react with aliphatic aldehydes and ketones to give imines as shown in section 4.1, but further reaction usually takes place. Treatment of formaldehyde with ammonia, for example, gives hexamethylene tetramine.

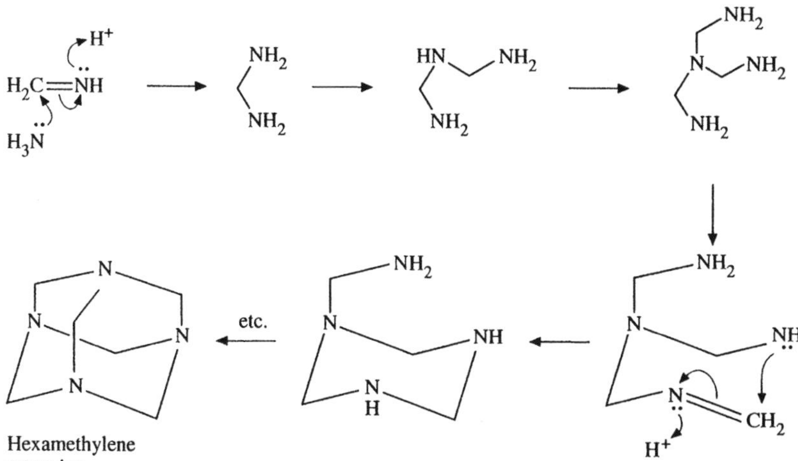

Hexamethylene tetramine

Aromatic aldehydes and ketones give isolable but still quite reactive imines with primary aliphatic amines. Their greater stability is due to the conjugation of the carbon–nitrogen double bond with the aromatic ring.

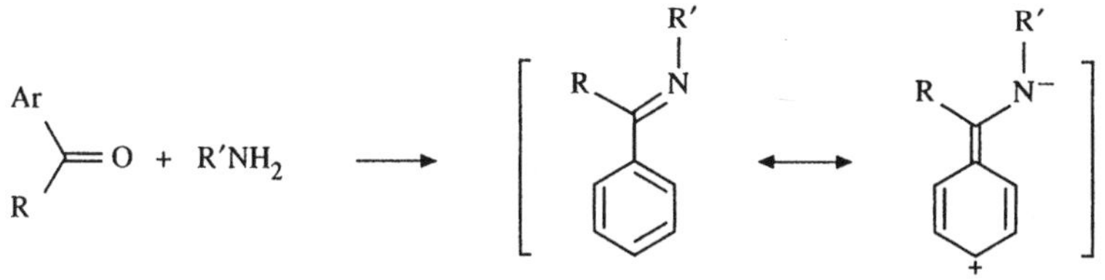

If an intramolecular hydrogen bond is present, as in imines derived from salicylaldehyde, this further stabilizes the system.

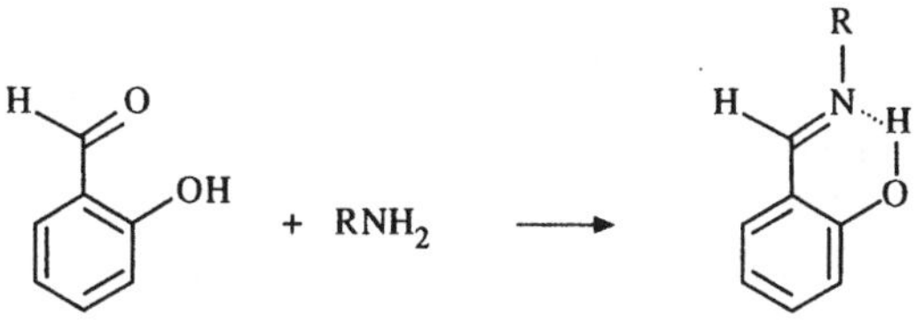

Salicylaldehyde

Aromatic primary amines also condense to give imines (sometimes called anils) which are stabilized by delocalization.

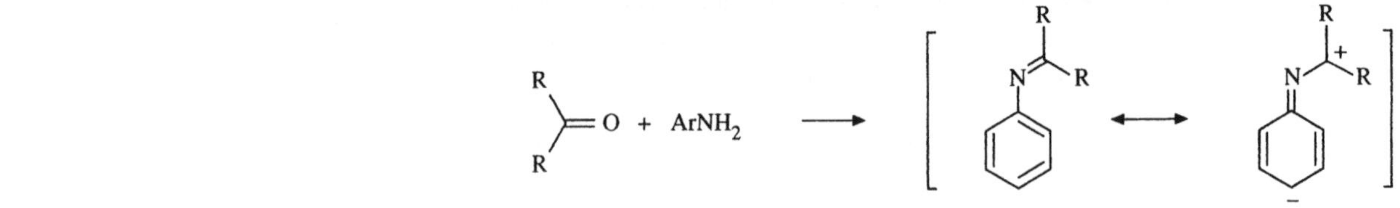

An imine is a planar polarized functionality like a carbonyl group. It reacts analogously towards nucleophiles, especially if acidic catalysis is available to assist by pulling electron density through nitrogen.

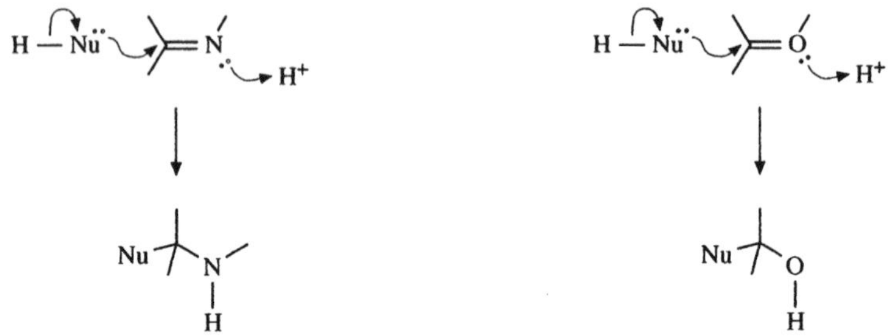

Imines are, for example, quite easily hydrolysed by reversal of the equilibria involved in their formation, and the reactions of aldehydes and ketones with cyanide, bisulfite, complex metal hydrides, Grignard reagents, etc all have parallels in the chemistry of imines. Of particular value are the so-called reductive alkylation procedures, in which ammonia or an amine and an aldehyde or ketone are brought together in reducing conditions, so that the imine which is formed is converted immediately to an *N*-alkyl derivative.

The Leuckart reaction, for example, yields amines when aldehydes or ketones are heated with ammonium formate. The reductive stage involves hydride transfer.

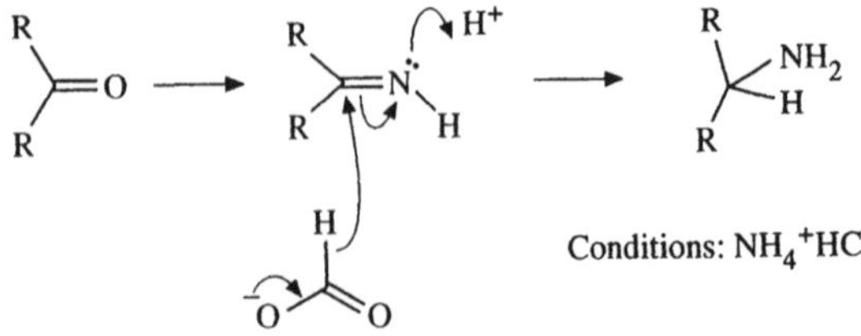

Conditions: $NH_4^+HCO_2^-/\Delta$

The Eschweiler–Clarke procedure for the *N*-methylation of secondary amines is closely related to the Leuckart reaction.

$$RNH_2 + H_2CO + HCO_2H \xrightarrow{\Delta} RNMe_2, \text{ via } RNHMe$$

Here there is a very electrophilic iminium ion intermediate, which is reduced by hydride transfer from formate.

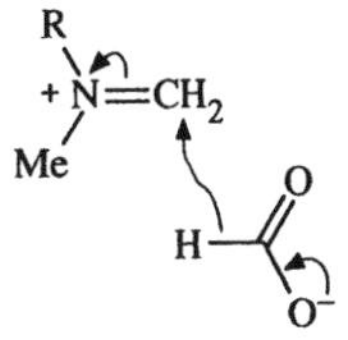

Sodium cyanoborohydride is an elegant reagent for reductive alkylation. It works as a reducing agent like $NaBH_4$, but the cyanide ligand gives the complex anion greater kinetic stability, and it can be used in the pH range 5–6, which are favourable conditions for imine formation. Mixing an aldehyde or ketone with an appropriate alkylamine at that pH in the presence of sodium cyanoborohydride thus gives an alkylated amine directly, in a single operation, as shown below.

$$PhCHO + EtNH_2 \longrightarrow PhCH_2NHEt, \text{ via } PhCH=NEt$$

Conditions: $NaBH_3CN/MeOH/pH5-6/20\,°C$

Here, as with all reductive alkylation procedures applied to primary amines, the secondary amine produced may react further with the starting carbonyl compound, to give, after reduction of the iminium ion, a tertiary amine. Excess ammonia or amine must be used if conversion is to be held back to the secondary amine stage.

4.4   Enamines

If an aldehyde or ketone which has α-CH reacts with a secondary amine, the iminium ion produced sheds that proton reversibly, giving an enamine.

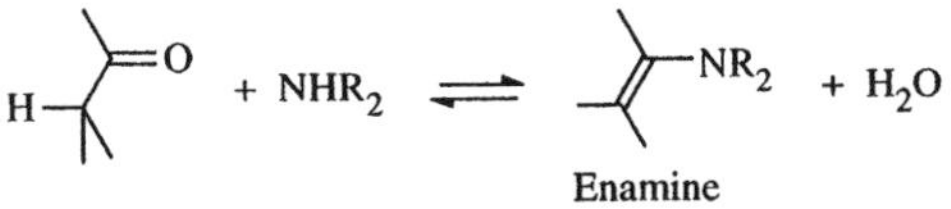

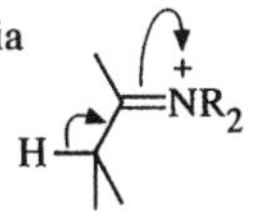

Such equilibria can be driven to the right by removal of water, and the resulting enamines can be isolated. Their chemical reactivity towards electrophiles is analogous to that of enols.

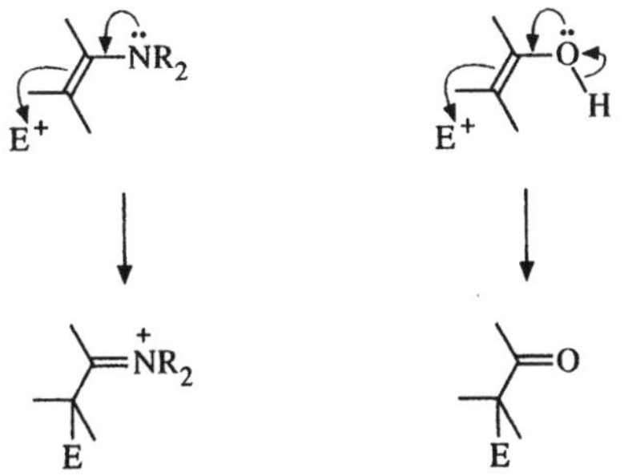

4.5   The Wolff–Kishner reaction

In this procedure, as modified by Huang Minlon, aldehydes and ketones are reduced to alkanes by heating with hydrazine and strong alkali in diethylene or triethylene glycol, as in the marginal example. The high-

$$PhCOEt \longrightarrow PhCH_2Et$$

Conditions: KOH/NH_2NH_2 in triethylene glycol at $180\,°C$

boiling (b.p. 285 °C) solvent enables high temperatures to be attained. Water is driven off as the reaction proceeds. The hydrazone which is formed under these conditions tautomerizes before loss of N_2 to give a carbanion, and thence the alkane product.

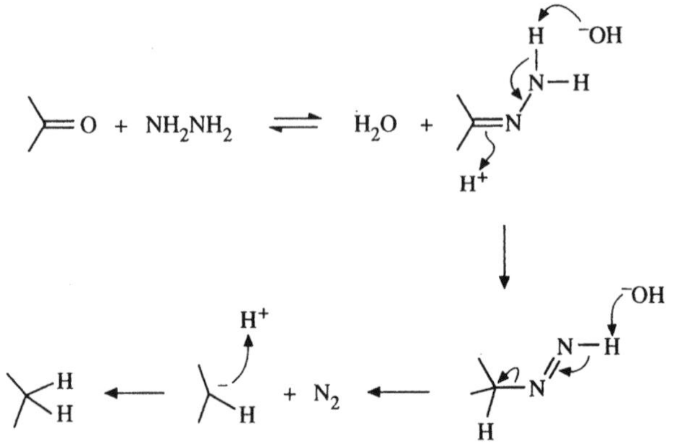

The carbanion intermediate is remarkable because it is not stabilized by delocalization. The elevated temperatures obviously help the formation of this high-energy species, but the principal driving force is probably the release of N_2, which is the most stable diatomic molecule known.

4.6 Transamination

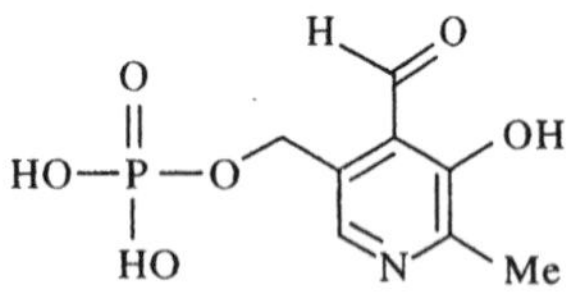

Pyridoxal 5-phosphate

Imine formation is also at the heart of a number of important biochemical processes involving pyridoxal-5-phosphate, including transamination, a stage in the degradation of α-amino acids. Transamination results overall in loss of the amino group by transfer to α-ketoglutaric acid, giving an α-keto acid and glutamic acid.

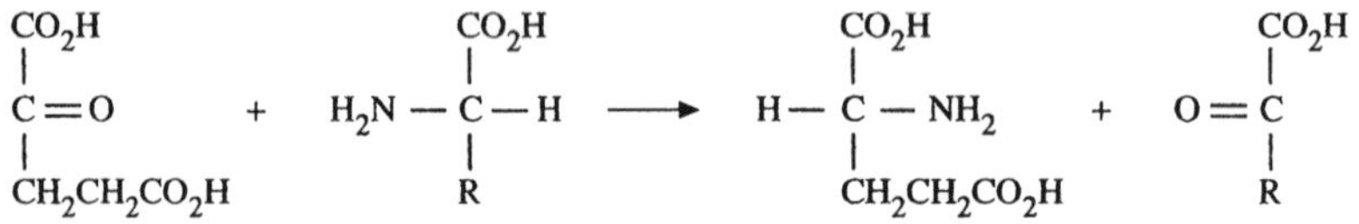

α-Ketoglutaric acid

Glutamic acid

The pyridoxal 5-phosphate, which is bound as an imine to an enzyme primary amine (lysine) side-chain, operates—all under enzymic mediation, of course—though successive imine exchange, isomerization, and hydrolysis.

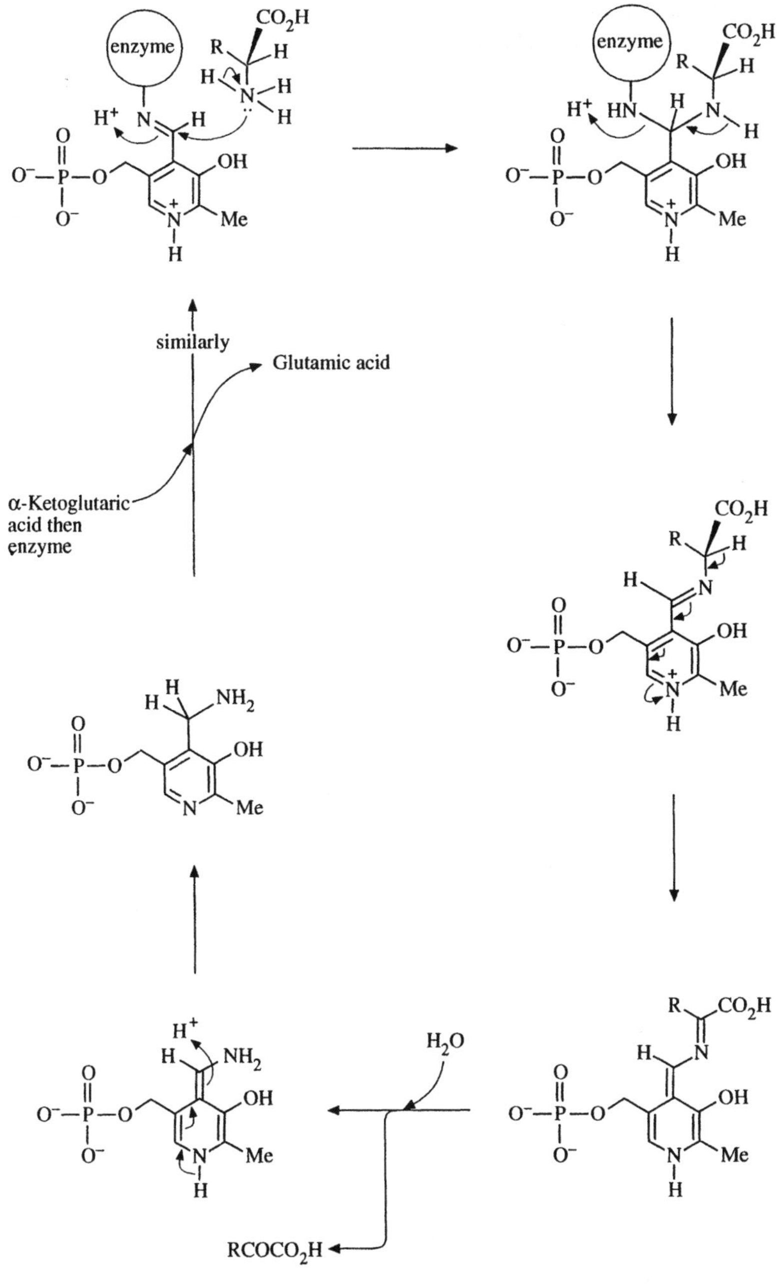
enzyme
similarly
Glutamic acid
α-Ketoglutaric acid then enzyme
H2O
RCOCO2H

5 Reactions of nucleophiles with carboxylic acid esters

5.1 Introduction

The electrophilic reactivity of the carbonyl carbon in esters is less than it is in aldehydes and ketones, because it is attenuated by the second oxygen atom.

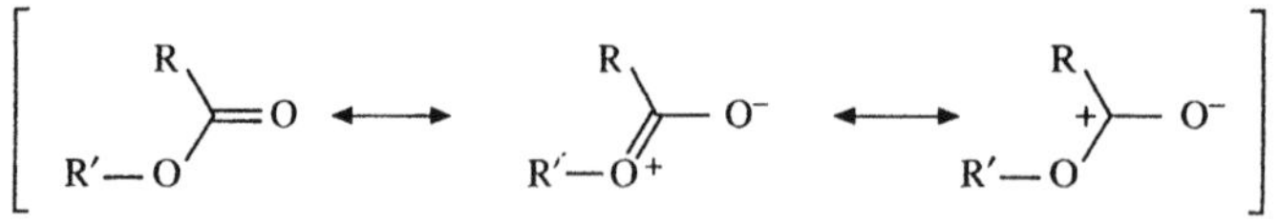

The overall outcome of the commonest pathway for reaction with anionic nucleophiles is quite different, but the first stage, nucleophilic addition, with sp^2 to sp^3 rehybridization, is the same.

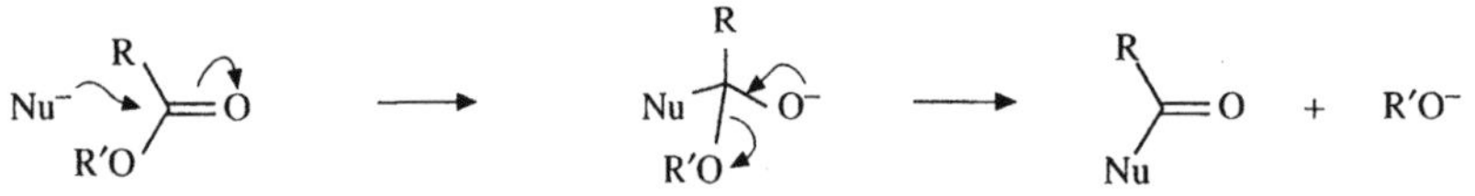

This process (on which there are many variants, as we shall see) is an acyl transfer reaction, in which the RCO– group is passed from an alcohol oxygen to a nucleophile, liberating the alcohol. Alternatively, less commonly, the nucleophile may react with the carbon of the alcohol moiety, in an S_N1 or S_N2 fashion.

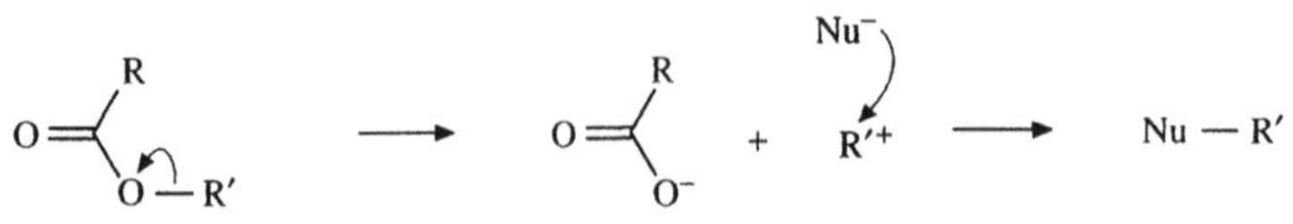

5.2 Hydrolysis

Introduction

When an alkyl ester of a carboxylic acid is hydrolysed, that is to say is cleaved into its component alcohol and carboxylic acid by water, it is not obvious by inspection whether the substrate bond which is broken is the acyl–oxygen bond or the alkyl–oxygen bond. Cleavage at (a) or (b) would yield the same products. An isotopic labelling experiment is necessary to resolve the ambiguity. Thus the hydrolysis of a simple methyl ester which has an ^{18}O-labelled saturated oxygen gives methanol containing all the ^{18}O: exclusive acyl–oxygen fission takes place in this case.

$$H_2O \ + \quad \begin{array}{c} R \\ \diagdown \\ Me - {}^{18}O \end{array} \!\!\! = O \qquad \longrightarrow \qquad Me^{18}OH + RCO_2H$$

Alkyl–oxygen fission can be similarly proved to occur when *t*-butyl esters are hydrolysed under acidic conditions.

$$H_2O \ + \quad \begin{array}{c} R \\ \diagdown \\ Bu^t \dashv {}^{18}O \end{array} \!\!\! = O \qquad \longrightarrow \qquad Bu^tOH \ + \ RC^{18}O_2H$$

Ester hydrolyses may be acid- (A) or base- (B) induced (hydrolysis in completely neutral conditions is slow in simple cases), may go with acyl–oxygen (AC) or alkyl–oxygen (AL) fission, and may have uni- (1) or bi- (2) molecular rate-determining steps. We shall consider only the three most common mechanisms, which are designated $B_{AC}2$, $A_{AC}2$, and $A_{AL}1$, and a less general mechanism which is designated $A_{AC}1$, although these do not by any means exhaust the field.

The $B_{AC}2$ mechanism

On treatment with aqueous alkali, methyl and ethyl esters are hydrolysed irreversibly via a tetrahedral adduct of substrate and OH^-, the formation of which is rate-determining. The overall process is irreversible, because the carboxylic acid product is fully ionized.

The $B_{AC}2$ mechanism

Hydrolysis also takes place under these conditions with esters of secondary alcohols, but the mechanism is sensitive to steric hindrance, and *t*-butyl esters are quite inert to alkali and other nucleophilic reagents. Electron-withdrawing substituents on either side of the ester functionality favour the formation of the tetrahedral intermediate, and accelerate the reaction. The rate constant for the $B_{AC}2$ hydrolysis of ethyl benzoate, for example, is increased about a hundredfold by placing a nitro group at the *p*-position of the aromatic ring. Similarly, in a series of phenyl acetates, which are as a class all subject to facile $B_{AC}2$ hydrolysis, electron-withdrawing substituents in the aromatic moiety accelerate the reaction markedly.

The hydrolysis of alkyl esters by alkali is often called saponification, a term which originally had the more limited meaning of soap-production by the action of alkali on oils and fats, which are mixtures of triesters of glycerol with long chain carboxylic acids (fatty acids), such as palmitic, stearic, and oleic acids.

$$
\begin{array}{c}
RCO_2CH_2 \\
| \\
RCO_2CH \\
| \\
RCO_2CH_2
\end{array}
+\; 3NaOH \;
\xrightarrow[\;H_2O\;]{\Delta}\;
\begin{array}{c}
HOCH_2 \\
| \\
HOCH \\
| \\
HOCH_2
\end{array}
+\; 3RCO_2Na
$$

Fat   Glycerol / Soap

R = $CH_3(CH_2)_{14}$, palmitic acid; R=$CH_3(CH_2)_{16}$, stearic acid;
R = $CH_3(CH_2)_7CH=CH(CH_2)_7$, oleic acid

The $A_{AC}2$ mechanism

Methyl and ethyl esters are also hydrolysed by aqueous acid via a tetrahedral intermediate, but this mechanism is reversible.

The $A_{AC}2$ mechanism

$RCO_2H + MeOH \rightleftharpoons$

If, instead of ester and excess water with a catalytic amount of strong acid, we begin with carboxylic acid and excess alcohol and a catalytic amount of strong acid, we can drive the whole system of equilibria backwards, to completion if we remove water as it is formed. Used in that reverse sense, the process is called esterification. Whether going forwards to hydrolyse an ester, or backwards to esterify an acid, the $A_{AC}2$ mechanism is, like the $B_{AC}2$ mechanism, sensitive to steric hindrance; but the response of the two pathways to electronic effects is quite different. Whereas the rate of a $B_{AC}2$ reaction is greatly increased by electron-withdrawing substituents near the reaction centre (see above), $A_{AC}2$ reactions are insensitive to electronic effects. This is not because the effects are inoperative, but rather because they operate in an opposing sense on different stages and balance so far as the overall rate is concerned. The overall rate of an $A_{AC}2$ reaction depends on the electrophilic reactivity of the carbonyl group in the protonated ester intermediate. This is enhanced by an electron-withdrawing substituent on either side, as with the $B_{AC}2$ mechanism. However, the overall rate also depends on the concentration of the protonated ester intermediate, and thus in turn on the basicity of the ester functionality, which is diminished by an electron-withdrawing substituent on either side. The rate constants for the $A_{AC}2$ hydrolysis of ethyl *p*-nitrobenzoate and benzoate are consequently about the same.

The $A_{AL}1$ mechanism

Esters of tertiary alcohols such as *t*-butanol are inert to alkali, as already mentioned, because of the very unfavourable steric interactions which would attend the formation of a $B_{AC}2$ tetrahedral adduct. For the same reason the $A_{AC}2$ pathway is not open to *t*-butyl esters under acidic conditions. Rapid hydrolysis nevertheless takes place under acidic conditions, but by alkyl–oxygen fission through a mechanism which is termed $A_{AL}1$, although it is in fact an S_N1 reaction of the protonated substrate.

The $A_{AL}1$ mechanism

This is a feasible process because a relatively stable $(CH_3)_3C^+$ ion is formed in the key step. Alkyl–oxygen fission in the same way with a methyl primary ester would give a CH_3^+ ion, an intermediate of prohibitively high energy, so it does not occur.

t-Butyl esters can be prepared by treating carboxylic acids with excess *t*-butanol in the presence of a strong acid catalyst, when $(CH_3)_3C^+$ ions formed by the *t*-butanol are caught by the carboxylic acid. This is simply the $A_{AL}1$ hydrolysis in reverse. The most convenient general method in practice, however, is to react the carboxylic acid with excess isobutene and a trace of strong acid, when $(CH_3)_3C^+$ ions formed by protonation of the isobutene are trapped by the carboxylic acid.

$$RCO_2H + Me_2C{=}CH_2 \longrightarrow RCO_2Bu^t$$

Conditions: excess $Me_2C{=}CH_2$/
trace H_2SO_4/20 °C/pressure bottle

The $A_{AC}1$ mechanism

This is a mechanism of much more limited scope than the three outlined above. Methyl mesitoate has a carbonyl group which is sterically hindered by the *o*-methyl groups on the aromatic ring. The $B_{AC}2$ and $A_{AC}2$ routes for hydrolysis are thus closed off. Since it is a methyl ester, the $A_{AL}1$ route is not open either. $B_{AL}2$ ($= S_N2$) cleavage is possible in theory, but in practice is very slow. However if this ester is dissolved in

Methyl mesitoate

concentrated sulfuric acid and the resulting solution is quenched with water, quantitative conversion to the free acid and alcohol ensues. In this reaction, the protonated ester functionality springs apart to give an acylium ion, with relief of steric compression because the acylium ion is linear, as drawn. Although stable in concentrated sulfuric acid (a medium lacking good nucleophiles), on adding water the acylium ion is trapped to give the carboxylic acid.

The $A_{AC}1$ mechanism

By quenching a solution of the carboxylic acid in concentrated sulfuric acid with excess methanol, the reaction may be operated backwards, as an esterification.

Asprin

pH dependence of the rate of hydrolysis of aspirin.

Intramolecular and enzymic catalysis

Ester hydrolysis can, if the appropriate structural features are present, be greatly accelerated by intramolecular catalysis. Aspirin, for example, is hydrolysed in the pH range 3–9 much more rapidly than its *p*-isomer—which, in the absence of special effects, might have been expected to be the more reactive because it lacks the *o*-steric hindrance of aspirin. It is clear that in this case the *o*-substituent actually helps the hydrolysis. The way in which the rate of hydrolysis depends on pH reveals that the sharp increase in rate between pH 3 and pH 4 is in step with the dissociation of the carboxyl group (pK_a 3.5), implying that the hyper-reactive species is the carboxylate anion. Without additional evidence there are two possible mechanisms to consider for its hydrolysis. Intramolecular nucleophilic catalysis has been ruled out by running the hydrolysis of aspirin in $H_2^{18}O$: no ^{18}O was incorporated into the salicylic acid produced, some would have been (by cleavage a) if an anhydride intermediate had been involved. This result is, however, consistent with intramolecular general base catalysis, which involves carbon–oxygen bond formation by

(a) Intramolecular nucleophilic catalysis

$$+H^+$$

This intermediate is an anhydride, which hydrolyses rapidly by water attack at either CO group (cleavage a or b) giving salicylic and acetic acids in either case.

(b) Intramolecular general base catalysis.

$$+H^+$$

Salicylic acid

$$+ \quad CH_3CO_2H$$

attack at the acetyl carbonyl group exclusively. The balance between the two pathways is fine: application of the same criterion to 3,5-dinitroaspirin shows that its hydrolysis does involve intramolecular nucleophilic catalysis.

Intramolecular catalysis is effective because helper groups are held near the reaction site, poised ready to guide the chemistry more swiftly and smoothly than intermolecular interactions dependent on chance encounter could. There is an enormous jump of complexity and subtlety between simple cases of intramolecular catalysis, like the hydrolysis of aspirin, and enzymic catalysis, but the bottom line is the same: enzymic catalysis depends on catalytic functionalities which are able to operate with delicacy and precision in the temporary enzyme–substrate assembly, because they are correctly placed for optimal effectiveness. Nothing is left to chance.

The chymotrypsin hydrolysis of *p*-nitrophenyl acetate is a relatively simple example. Chymotrypsin is a digestive enzyme whose natural function is the degradation of food protein by catalysis of peptide bond hydrolysis. Like all enzymes, it is a macromolecular protein designed by Nature to have an active site sculpted to fit the substrate and bring it into reaction with the catalytic groups. *p*-Nitrophenyl acetate is not a natural or even a reactive substrate for chymotrypsin, but it has served as an experimentally convenient model. When it is mixed in solution with the enzyme, there is an immediate release of one mole of *p*-nitrophenol per mole of chymotrypsin. This corresponds to the fast formation of an acetylenzyme; slower hydrolysis of this covalent intermediate completes the reaction and regenerates catalytic enzyme. After the initial burst of *p*-nitrophenol, the rate of its release is steady and limited by the rate at which the enzyme is regenerated.

The functional group in the enzyme active site which is acetylated and deacetylated is a primary hydroxyl group provided by a serine side-chain:

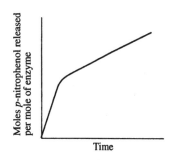

Chymotrypsin-catalysed hydrolysis of *p*-nitrophenyl acetate.

both stages are subject to general base catalysis by an adjacent heterocyclic base (imidazole) provided by a histidine side-chain. The two stages can be represented as shown below.

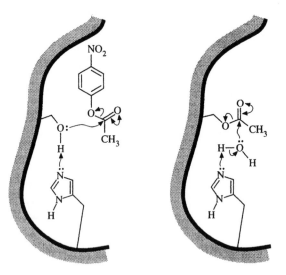

Acetylation of an OH group in the active site   Deacetylation of the acetylenzyme

5.3   Transesterification

Some of the mechanistic pathways outlined in Section 5.2 for ester hydrolysis can also be used to convert one ester to another. Such interconversions, which are also encountered as side reactions or unintended complications from manipulations in alcoholic solvents, are called transesterification reactions.

A methyl ester will be converted to an ethyl ester if it is dissolved in ethanol containing sodium ethoxide ($B_{AC}2$) or an acid catalyst ($A_{AC}2$); a *t*-butyl ester will not be affected by ethanol containing sodium ethoxide, but in ethanol with an acid catalyst conversion to an ethyl ester will take place ($A_{AL}1$ cleavage followed by $A_{AC}2$ esterification); and methyl mesitoate is converted to the corresponding ethyl ester by dissolving it in concentrated sulfuric acid and quenching with ethanol ($A_{AC}1$).

5.4   Ammonolysis and related reactions

Ammonia and its derivatives—alkylamines, hydroxylamine, hydrazine, etc.—all react with simple esters (except those of tertiary alcohols) under neutral or basic conditions, by mechanisms involving tetrahedral adducts which are broadly similar to the $B_{AC}2$ and $A_{AC}2$ mechanisms of hydrolysis.

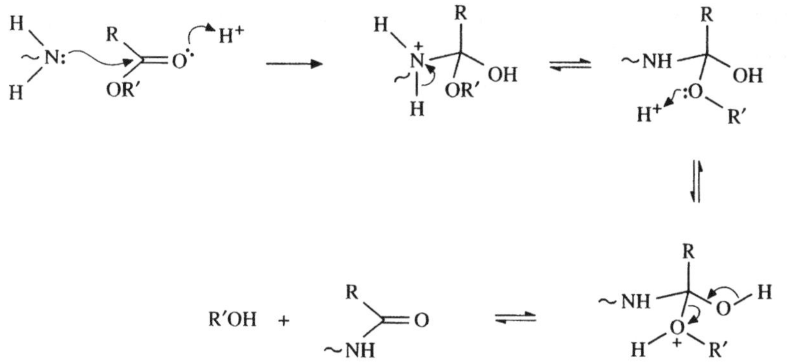

There are many subtleties of mechanism which are beyond our scope, but note that NH_2^-, RNH^- and R_2N^- are all such strong bases that significant concentrations cannot exist in aqueous or alcoholic solutions; and in strongly acidic solutions ammonia and alkylamines are fully protonated, and therefore not nucleophilic. The reaction of ammonia itself with ethyl acetate is rather slow.

$$CH_3CO_2Et + NH_3 \longrightarrow CH_3CONH_2 + EtOH$$

Conditions: conc. aq. ammonia/several days/20 °C

For reactions of amino compounds with carboxylic acid derivatives to give peptides, see Primer 7.

Alkylamines are even slower to react: the electron-releasing effect of an alkyl group raises the intrinsic reactivity of the nitrogen slightly, but this is more than offset by the additional steric hindrance, to which, as we have seen, all reactions of this class are sensitive. As with hydrolysis, electron-withdrawing substituents on either side of the ester function greatly enhance its reactivity. Ethyl chloroacetate and *p*-nitrophenyl acetate, for example, are both much more reactive towards ammonia than ethyl acetate is.

Although hydroxylamine and hydrazine are both weaker bases than ammonia, they are better nucleophiles in acyl transfer reactions. They react with esters to give hydroxamic acids and hydrazides, respectively; take special note that they do not yield oximes or hydrazones, as unthinking comparison with the behaviour of aldehydes and ketones might suggest.

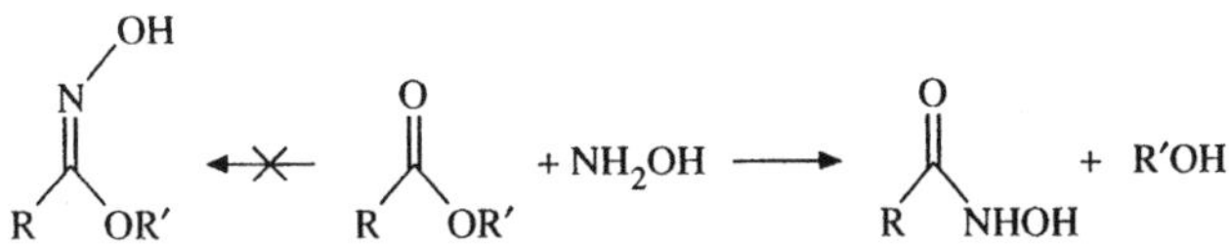

A hydroxamic acid

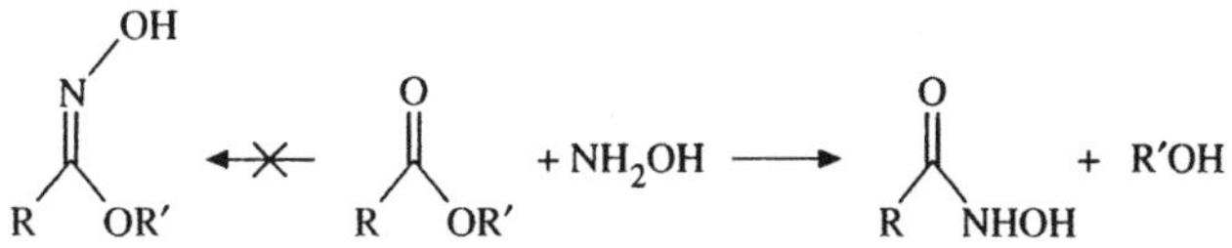

A hydroxamic acid

5.5   LiAlH₄, NaBH₄ and DIBAL

LiAlH$_4$ reacts with most kinds of esters—except, again, those of tertiary alcohols—by addition to give a tetrahedral intermediate, which collapses to give an aldehyde and an alcohol; since the aldehyde is more electrophilic than the starting ester, it is rapidly reduced. The overall result, after working up with aqueous acid, is reductive cleavage of the ester to two alcohols.

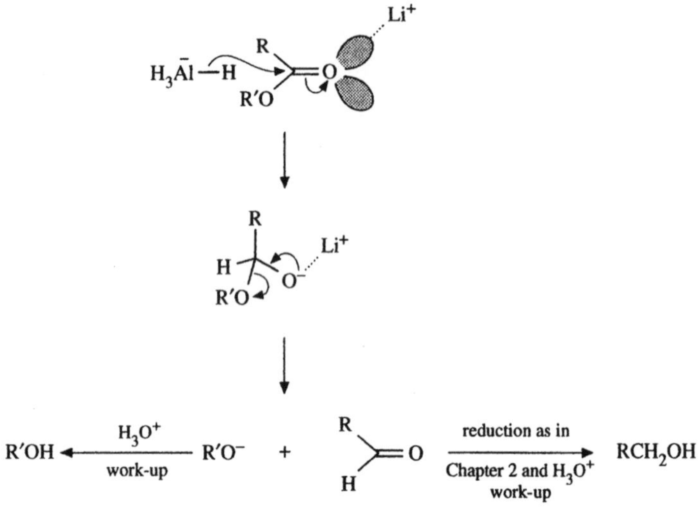

Simple esters such as ethyl acetate are not, however, reduced by NaBH$_4$, exemplifying the low electrophilicity of esters compared to aldehydes and ketones. But the bald statement which appears in some texts to the effect that NaBH$_4$ does not reduce esters is untrue. It is a matter of degree, not kind, and ester groups with electron-withdrawing substituents nearby are in fact reduced by NaBH$_4$ without difficulty. Lithium borohydride lies between LiAlH$_4$ and NaBH$_4$ in its reactivity towards carbonyl groups, illustrating the importance of the metal ion in these complex metal hydride reducing agents: it reduces simple esters.

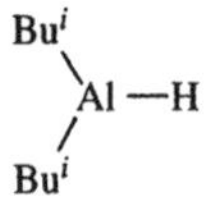

DIBAL—diisobutylaluminium hydride—also reduces simple esters at room temperature in a two-stage process giving two alcohols.

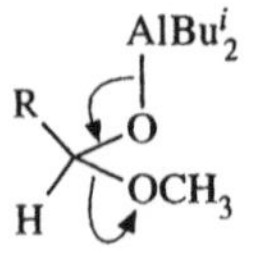

DIBAL

Intermediate collapses to give
aldehyde which is reduced

$$RCO_2CH_3 \longrightarrow RCH_2OH + CH_3OH$$

Conditions: DIBAL/20 °C, then H$_3$O$^+$ work-up

However, in this case the collapse of the aluminium alkoxide intermediate produced by the first stage is slow enough for the reduction to be halted there at −78 °C: on aqueous work-up an aldehyde is obtained.

$$RCO_2CH_3 \longrightarrow RCHO + CH_3OH$$

Conditions: DIBAL/–78 °C, then H_3O^+ work-up

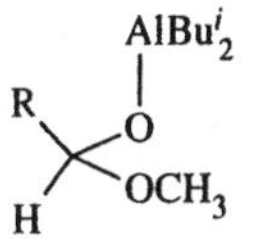

Intermediate survives to
give aldehyde on work-up

5.6 Grignard reagents

Grignard reagents react with esters in two stages (cf. their reduction by $LiAlH_4$) giving two alcohols after work-up, as in the example below.

$$2PhMgBr + PhCO_2Et \longrightarrow Ph_3COH + EtOH$$

Conditions: Et_2O as solvent; work-up with H_3O^+

Structure correlates with reactivity predictably: electron-withdrawing substituents pulling on the ester facilitate and adjacent bulky groups inhibit (*t*-butyl esters do not react).

Problems

1. $CH_3COCH_2CO_2Et$ reacts with hydrazine to give a product $C_4H_6N_2O$. Suggest a structure for this product and explain how it is formed.

2. Benzyl esters RCO_2CH_2Ph react with anhydrous HBr to give RCO_2H. What is the other product and how is it formed?

3.

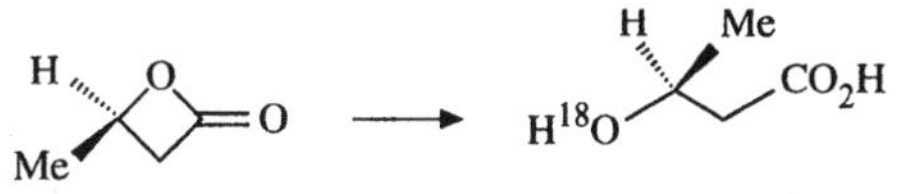

Conditions: i, $H_2^{18}O$, mild acid

6 Reactions of nucleophiles with other carboxylic acid derivatives

The esters which were considered in the previous chapter lie roughly in the middle of a reactivity series in which the ease of nucleophilic attack at the carbonyl carbon decreases from left to right.

$$RCO-Cl > RCO-OCOR > RCO-OR > RCO-NR_2 > RCO-OH$$

Thiol esters, which are mentioned briefly in Section 9.4, are between ordinary esters and carboxylic acid anhydrides in reactivity.

6.1 The carboxylic acids themselves

The carboxylic acids themselves are the least reactive of the above series to, say, Grignard reagents, because they are converted to carboxylate complexes on which nucleophilic attack is very difficult.

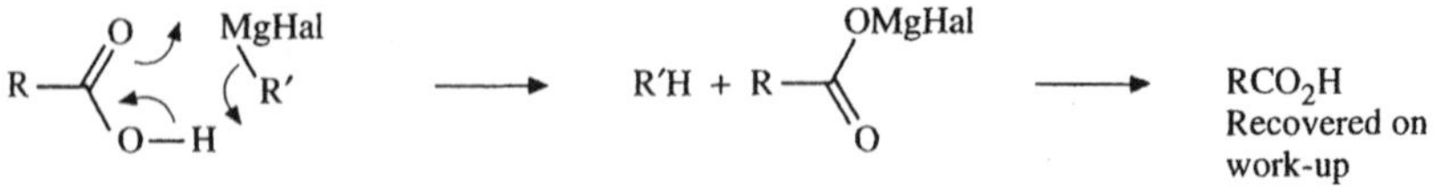

In fact, even with excess Grignard reagent, no further reaction takes place, and the carboxylic acid is recovered on acidification. The more reactive organolithium reagents, however, are able to carry things through, and the carboxylate produced by the first equivalent of reagent reacts with a second, giving an adduct which survives to yield a ketone on work-up.

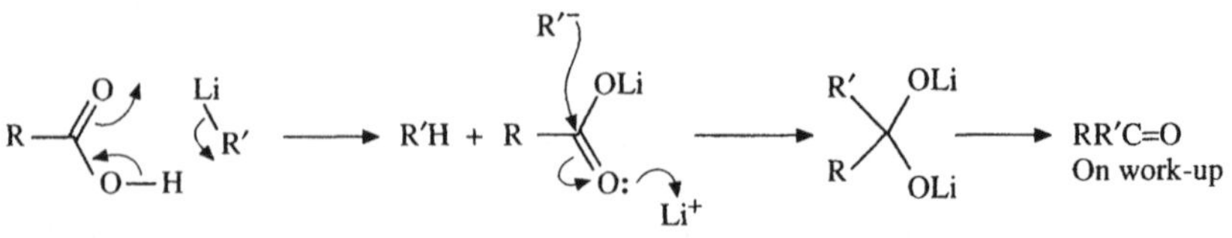

Ph$_2$CHCO$_2$H $\longrightarrow$ Ph$_2$CHCH$_2$OH

Conditions: i, LiAlH$_4$/Et$_2$O/reflux; ii, dil. H$_2$SO$_4$ work-up

LiAlH$_4$ reduces carboxylic acids to primary alcohols, via a carboxylate complex again, as in the marginal example, but NaBH$_4$ is not reactive enough to do this.

Ammonia and amines simply form salts with carboxylic acids, and these are stable except at high temperatures. To achieve carbon–nitrogen bond formation, it is necessary first to activate the carboxyl group by replacing the OH with a better leaving group, such as Cl, which can be brought about indirectly by reaction with an electrophilic reagent. Thionyl chloride is a suitable reagent: the carbon–chlorine bond forms after attack at oxygen.

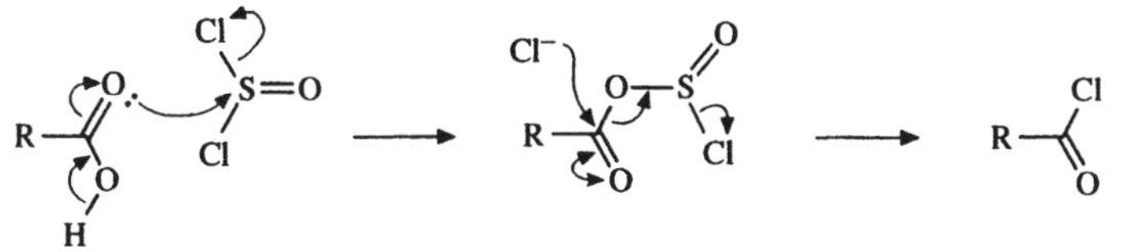

Acyl chlorides react very easily with ammonia and amines: see Section 6.4.

Although carboxylic acids are unreactive towards nucleophilic reagents, they are reduced to primary alcohols very easily with borane followed by aqueous work-up. This is because the primary events with this reagent involve it acting as an electrophile. The reaction is mechanistically complex, but the essence of what happens can be outlined as shown below.

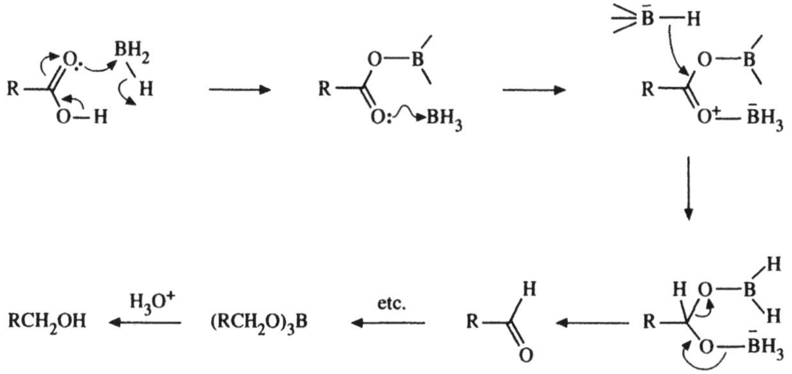

The complexity of the reaction is obvious if we reflect that the borane reagent is actually a dimer in the gas phase, which might react as such in solution; that the reaction might involve a mono-, di-, or tri-acyloxyoborane; that once underway, there are likely to be several different Lewis acids in the field of play; that there are several candidates for the complex which transfers hydride to make the new carbon–hydrogen bonds; that whether one draws intramolecular or intermolecular transition states for some stages is more a matter of taste than evidence; and so on. It is hardly surprising that different texts offer apparently different mechanisms (compare Norman and Coxon with Kemp and Vellaccio), and that some texts baulk at the prospect and say nothing. However, the student need not be confused; the niceties make no difference to the outcome or to the main point, which is that substrate reactivity towards borane depends not on the electrophilicity of the carbonyl carbon, but on the Lewis basicity of the carbonyl oxygen.

Ketone-like

Amine-like

Orthogonal

6.2 Amides

The electrophilic reactivity of the carbonyl carbon in amides is low, because of interaction with the adjacent nitrogen atom, which is essentially nonbasic for the same reason. The nitrogen atom is sp^2-hybridised, and the whole system has a strong preference for planarity, because only then are the three *p*-orbitals aligned to overlap. This restraint is of far-reaching significance in the structure of proteins, which are polyamides. If an amide group is artificially distorted by incorporating it into a bicyclic ring system, the normal conjugation cannot take place and the carbonyl group is ketone-like.

Amides are hydrolysed to carboxylic acids by aqueous alkali and strong aqueous acid, by $B_{AC}2$ and $A_{AC}2$ pathways like esters, but much more vigorous conditions are required—typically overnight reflux.

The reduction of amides with LiAlH$_4$ involves, in its first stage, nucleophilic addition to the carbonyl group, analogous to (albeit somewhat slower than) the reduction of esters. Whereas in the ester case at room temperature the tetrahedral adduct decomposes into two parts with cleavage of the molecular backbone (see Section 5.5), in the amide case the corresponding tetrahedral adduct decomposes with carbon–oxygen bond cleavage and preservation of the backbone. This is because $R_2'N^-$ is a very poor leaving group.

The H$^-$ here is not simply displaced, but reduces another CO group.

If LiAlH$_4$ reduction is performed at low temperature the decomposition of the tetrahedral adduct may be slow, and aqueous work-up gives an aldehyde. This sort of procedure only works well if there is also a structural feature in the amide which inhibits collapse of the adduct by making electron release by the nitrogen atom difficult.

Conditions: LiAlH$_4$/Et$_2$O/0 °C, then H$_3$O$^+$ work-up

Similarly, Grignard reagents react with *N,N*-disubstituted amides to give ketones, and with dimethylformamide to give aldehydes.

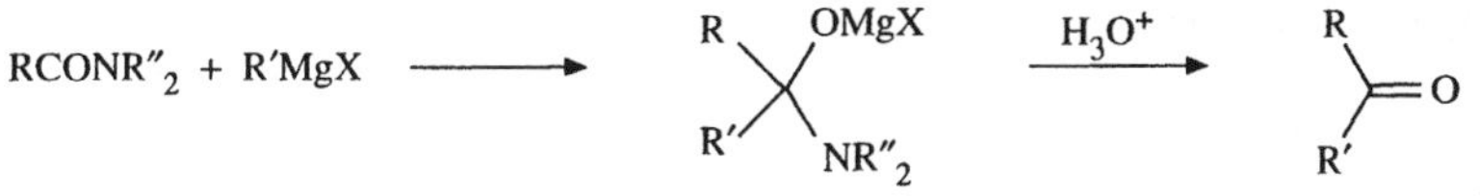

$$RCONR''_2 + R'MgX \longrightarrow$$

Adduct survivies
until work-up

$$\xrightarrow{H_3O^+}$$

Organolithium reagents react analogously with *N,N*-dialkyl amides but the tetrahedral adducts tend to collapse to ketones, resulting in further reaction. This difficulty can be completely overcome if the adduct is stabilized by chelation, as in the following example of the so-called Weinreb amide procedure.

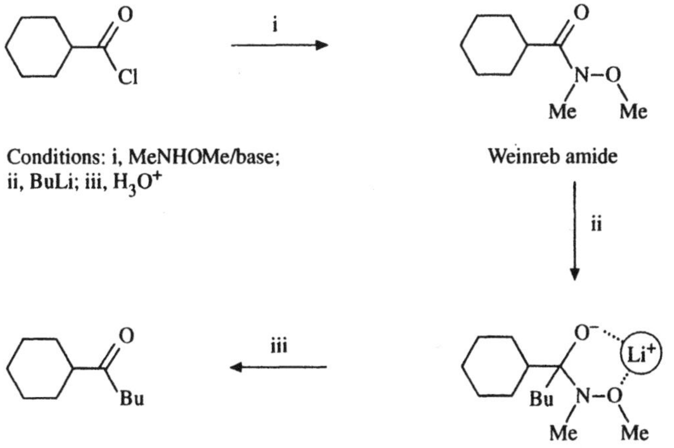

Conditions: i, MeNHOMe/base;
ii, BuLi; iii, H$_3$O$^+$

Weinreb amide

Adduct stabilized
by chelation

6.3   Carboxylic acid anhydrides

Carboxylic acid anhydrides are considerably more reactive towards nucleophiles than esters, because RCO_2^- is a much better leaving group than RO^-.

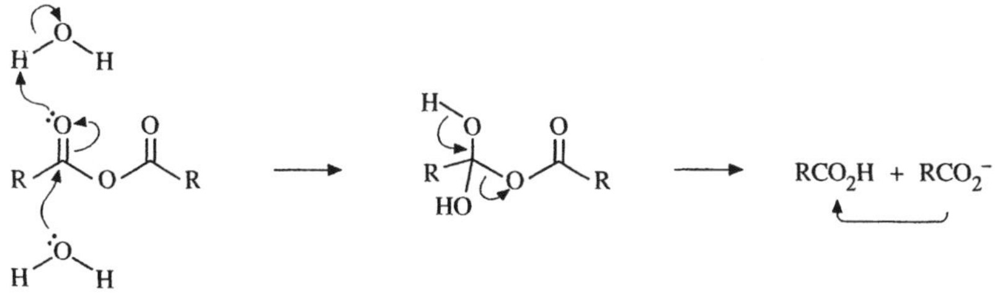

Hydrolysis and alcoholysis occurs easily, without the need for added acid or base; ammonia and amines react vigorously, and so do Grignard reagents and complex metal hydride reducing agents.

6.4   Acyl chlorides

Acyl chlorides, also called acid chlorides, are exceedingly and indeed sometimes violently reactive towards nucleophilic reagents. They fume in moist air because of facile hydrolysis to the carboxylic acid and HCl.

Alcohols react analogously, to give esters; carboxylic acids, and their salts, react to give anhydrides; and amines give amides, but excess amine or added base is required to take up the HCl which is produced. The Schotten–Baumann procedure (see below for an example) is a convenient way of performing the reaction in simple cases. It entails adding the acyl chloride to an efficiently agitated alkaline solution of the amine, which is maintained at a high enough pH to keep the amine in its free base nucleophilic form. Cooling is often necessary. The procedure works better for aroyl chlorides than alkanoyl chlorides, because the latter are so reactive that hydrolysis competes with aminolysis.

$$PhCOCl + PhNH_2 \longrightarrow PhCONHPh$$
Benzanilide

Schotten–Baumann procedure

Conditions: reactants shaken vigorously
with 10% aq. NaOH until the PhCOCl
is all consumed; product precipitates

Acyl chlorides react with Grignard reagents to give tertiary alcohols, because, as with esters, the intermediate ketone which is formed reacts further. Dialkyl cadmium reagents, however, while being insufficiently reactive to attack ketones or esters, are reactive enough to attack acyl chlorides, in a process which stops at the ketone stage. Similarly, LiAlH$_4$ reacts with ease in two stages to give a primary alcohol, but the less reactive reagent lithium tri-t-butoxyaluminium hydride is able to achieve only the first stage at low temperatures, providing a useful aldehyde synthesis.

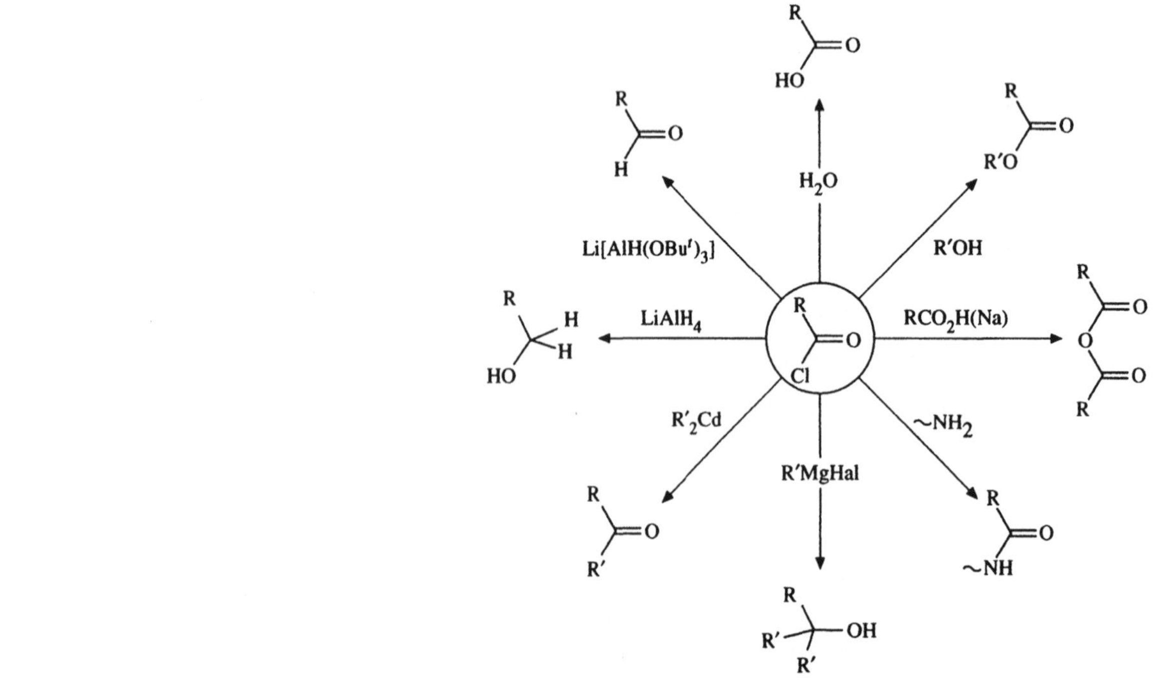

Reactions of acyl chlorides.

The high reactivity of acyl chlorides towards nucleophilic reagents arises because the chlorine atom is pulling electrons away from the carbonyl carbon, enhancing its electrophilicity. By the same token, the Lewis basicity of the carbonyl oxygen is diminished.

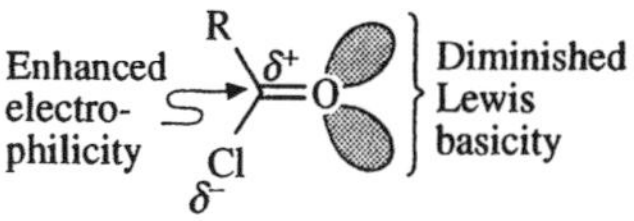

Borane, which reduces carbonyl groups by a mechanism involving as its first and critical stage a boron–oxygen Lewis acid–base interaction, does not react with acyl chlorides.

Although textbooks normally pigeon-hole the Friedel–Crafts acylation of electron-rich aromatic rings by acyl chlorides in the presence of Lewis acids as an electrophilic aromatic substitution, which is perfectly correct, the reaction also belongs here. The electrophile is an acylium ion, which is captured by the aromatic substrate acting as nucleophile, attacking at the carbon which belonged to the carbonyl group.

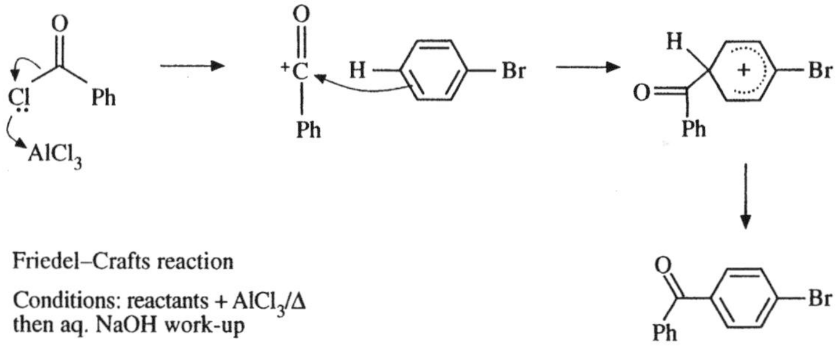

Enhanced electro-philicity δ^+ Diminished Lewis basicity δ^-

See Primer 4.

Friedel–Crafts reaction

Conditions: reactants + AlCl$_3$/Δ then aq. NaOH work-up

Looked at this way, the Friedel–Crafts acylation bears analogy with A$_{AC}$1 hydrolysis

Problems

1. Succesive treatment of PhCH$_2$CO$_2$H in anhydrous THF with Et$_3$N (1 equiv.), (CH$_3$)$_3$CCOCl (1 equiv.), and PhCH$_2$NH$_2$ (1 equiv.) gives PhCH$_2$CONHCH$_2$Ph. Explain.

2. The penicillin antibiotics such as amoxicillin are easily inactivated by hydrolysis. Where is the vulnerable part of the molecule? Explain.

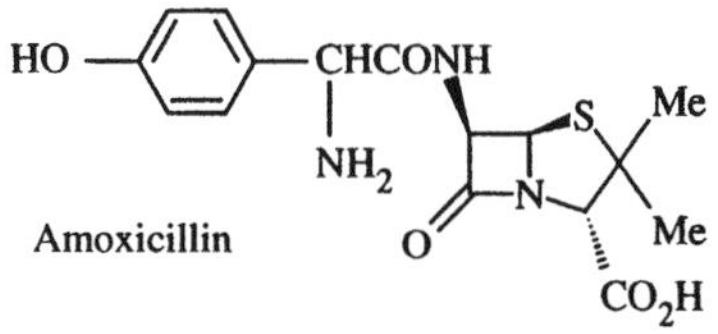

HO—⟨⟩—CHCONH   S   Me
         |
        NH$_2$ N
Amoxicillin   O   Me
               CO$_2$H

7 Enols and enolates

7.1 Enol and enolate formation

An aqueous solution of a simple ketone such as acetone is a surprisingly complex system. In addition to the reversible hydration of the keto form (see Section 2.2), it is in equilibrium with an enol tautomer and an enolate ion. At equilibrium, the concentrations of enol and enolate are very small in this case; the enol tautomer is of high energy compared to the keto tautomer, and the pK_a of acetone is about 20, compared to about 15–16 for water and alcohols. So the conjugate base derived from acetone (i.e. the enolate) can only exist at trace concentrations in aqueous or alcoholic solution.

Keto tautomer   Enolate ion

Keto tautomer   Enol tautomer

If two carbonyl groups flank a CH or CH_2 (which is thus α to two carbonyl groups) then the pKa of that CH or CH_2 is much higher (about 9 in the case of acetylacetone) than the CH_3 of acetone, because the enolate is stabilized by greater delocalization. The enolate can therefore exist at substantial concentrations in water and alcohols.

Acetylacetone   etc.

The corresponding enol is stabilized not only by delocalization, but also by the scope for the formation of an intramolecular hydrogen bond.

Ethyl acetoacetate, commonly called acetoacetic ester, is a similar case which has been thoroughly studied. At equilibrium in the pure neat liquid state at room temperature, the keto and enol forms are present in proportions of about 12:1. Careful fractional distillation, with rigorous exclusion of all possibility of acidic and basic catalysis of the equilibration, enables the two tautomeric forms to be separated. The pure enol gives an immediate bright red colour with ferric chloride, a classical colour test for phenols, which have OH attached to an unsaturated system, like enols; the pure keto tautomer does not give this colour test, except on standing. The enol reacts instantly with bromine; the keto form reacts only slowly, but to give the same monobromo product. If either pure tautomer is allowed to stand with traces of acid or base, an identical mixture is produced which is indistinguishable physically and chemically from acetoacetic ester as ordinarily prepared.

In principle, all carbonyl compounds can form enols and enolates. Because the electronic factors which stabilize the corresponding enolates tend to stabilize enols similarly, the enol contents of carbonyl compounds correlate with the pK_a values of their α-CH groups. Some α-CH pK_a values are given in the table, together with some more familiar acids to provide points of reference.

Acetoacetic ester

Some approximate pK_a values

CH_3CO_2H	5
$CH_3COCH_2COCH_3$	9
$PhOH$	10
$CH_3COCH_2CO_2Et$	11
$CH_2(CO_2Et)_2$	13
H_2O	15
$EtOH, MeOH$	16–18
CH_3COCl	16
CH_3CHO	17
Bu^tH	19
CH_3COCH_3	20
CH_3CO_2Et	25
Ph_3CH	33
NH_3, R_2NH	35

The conjugate base of any acid in the table will react with any carbonyl compound above it, to generate the enolate, quantitatively if the pK_a difference is sufficient. Thus treatment of acetoacetic ester in ethanol with an equivalent amount of sodium ethoxide causes complete conversion to the enolate.

Conditions: NaOEt/EtOH

A simple ketone like acetone or cyclohexanone will be left in the unionized form under the same conditions—the enolate will be there, and if consumed more will be formed rather rapidly in the same way, although there will never be much of it. If instead of sodium ethoxide in ethanol we take a base derived from a secondary amine, such as LDA, in an unreactive aprotic solvent like tetrahydrofuran, then quantitative enolate formation occurs, as in the following example.

Conditions: LDA/THF

A certain amount of chemical commonsense is required here, however. Thus the similar pK_a values of acetyl chloride and ethanol might suggest that addition of acetyl chloride to sodium ethoxide in ethanol would result in an equilibrium being set up with the formation of substantial amounts of the enolate.

$$CH_3COCl + NaOEt/EtOH \longrightarrow {}^{-}CH_2COCl \ ?$$

In fact, if done incautiously on a significant scale, a violent and possibly catastrophic reaction (nucleophilic displacement of the chlorine) would take place.

7.2   Simple enol and enolate reactions

Enols and enolates both react with electrophiles (see Section 1.2). We shall consider here some examples of reactions where the electrophile is H^+, D^+, or Br_2. Cases where the reactive atom of the electrophile is carbon, so that a new carbon–carbon bond is formed are very important in synthesis: they are dealt with in the two closing chapters.

Racemization and epimerization of chiral ketones

Single enantiomers of ketones with chiral α-carbons are easily racemized under acidic or basic catalysis, provided α-CH is present. This takes place via the enol or enolate respectively, both of which are planar and can accept a proton at carbon on either side of the plane, with equal probability, giving a racemate.

Conditions: acid–base cat.

If, however, the two products of *C*-protonation are not of equal energy, they will not be formed with equal facility. Thus the enolate derived from the two decalin derivatives shown in the margin is protonated preferentially to give the less strained *trans*-isomer, providing a means of transforming *cis* to *trans*.

Deuterium exchange

Aldehydes and ketones with α-CH undergo exchange of all such hydrogen by deuterium in D_2O. As with the racemization of chiral ketones, the process is both acid- and base-catalysed and involves the enol or enolate, respectively.

$$CH_3CH_2COCH(CH_3)_2 \longrightarrow CH_3CD_2COCD(CH_3)_2$$

Conditions: D_2O/acid–base cat.

Only the α-positions are affected. The rates of α-deuterium exchange of carbonyl compounds correlate with their pK_a values i.e. β-dicarbonyl compounds react faster than simple aldehydes and ketones, which are in turn more susceptible than simple esters.

Bromination

Acetone reacts easily with bromine under both acidic and basic catalysis, to give monobromoacetone as the first-formed product in both cases.

Conditions: a, basic; b, acidic.

The rate of product formation depends on [acetone], but is independent of [Br_2]; the rates of chlorination, bromination, and iodination under the same conditions are the same. In more complex ketones, only the α-position is brominated. These observations all point to mechanisms involving rate-determining enol and enolate formation.

For preparative monobromination of ketones, acid-catalysed procedures work well. Thus acetophenone reacts with bromine in acetic acid to give an excellent yield of phenacyl bromide. The reaction, which is autocatalytic because the HBr coproduct catalyses enol formation, does not proceed beyond the monobromo stage, as the enol derived from the product is deactivated towards further electrophilic attack, by the electron-withdrawing effect of the bromine atom. The rates of base-catalysed brominations of carbonyl compounds follow the same trends as their acidities—very fast with β-dicarbonyl compounds, rapid with simple aldehydes and ketones, very slow with simple esters. From a preparative point of view, difficulty arises from the fact that if there is more than one enolizable hydrogen in the substrate, the monobromo product has a more acidic hydrogen than the substrate, so a second bromination takes place. In the case of a methyl ketone, all three hydrogens are replaced by

PhCOCH$_3$

↓

PhCOCH$_2$Br

Phenacyl bromide

Conditions: Br_2/AcOH

treatment with bromine in aqueous alkali, but then further reaction takes place with carbon–carbon bond cleavage, to give tribromomethane (bromoform). The same reaction occurs on treatment of methyl ketones with iodine and alkali, giving iodoform. The iodoform usually separates as a pale yellow crystalline solid with a characteristic odour, and the reaction is a classical test-tube method of identifying methyl ketones. It is specific to methyl ketones (and compounds which are converted to methyl ketones under the conditions used) because it is necessary for three halogens to have been introduced for the cleavage stage to be possible.

Conditions: Hal_2/OH^-

Although carboxylic acids, simple esters and amides do not undergo acid- or base-catalysed bromination because the respective enols and enolates do not form easily enough, the Hell–Volhard–Zelinsky bromination of carboxylic acids gives easy access to α-bromocarboxylic acid derivatives. In one popular variation of this procedure, a carboxylic acid is treated with bromine and a little red phosphorus. Some PBr_3 is formed, which converts the carboxylic acid to the corresponding acyl bromide; the crucial stage probably entails enolization of the acyl bromide and reaction of its enol with bromine; acyl bromide interchange with unmodified carboxylic acid (via an acid anhydride, which may also be brominated through its enol) allows the process to be repeated until all the starting acid is consumed. Aqueous work-up gives a good yield of α-bromocarboxylic acid.

Only the α-position of the carboxylic acid is brominated, and only one bromine is introduced, except under forcing conditions. There are several useful modifications of the Hell–Volhard–Zelinsky method. It is not essential for the PBr_3 to be formed *in situ*. It can be added in stoichiometric amount, and PCl_3 may be used, when the key intermediate is the acyl chloride. Alternatively, in a convenient preparation of α-bromoesters, a carboxylic acid may be converted separately to the acyl chloride before bromination and alcoholysis, as in the marginal example.

$$EtCH_2CO_2H \xrightarrow{\text{i–iii}} EtCHBrCO_2Me$$

Reagents: i, $SOCl_2$; ii, Br_2; iii, MeOH

Reactions with nitrogen electrophiles

Enols and, better, enolates couple with aryldiazonium cations to give, after tautomerization, arylhydrazones.

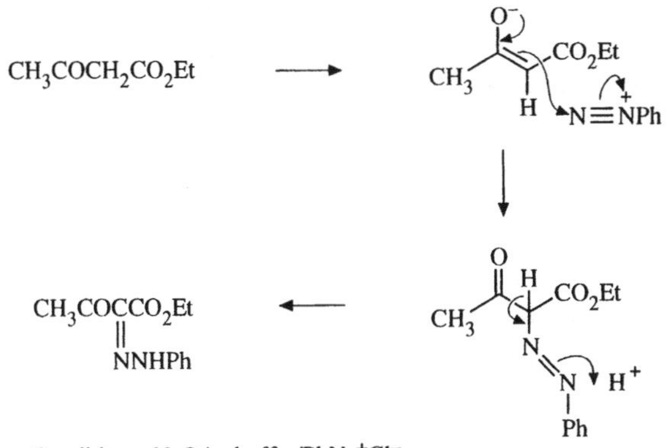

Conditions: NaOAc buffer/PhN$_2^+$Cl$^-$

Similarly, nitrous acid (i.e. acidified aqueous sodium nitrite solution, a source of NO$^+$ cations) reacts with enols to give, after tautomerization, oximes, providing through their hydrolysis a useful route to α-diketones.

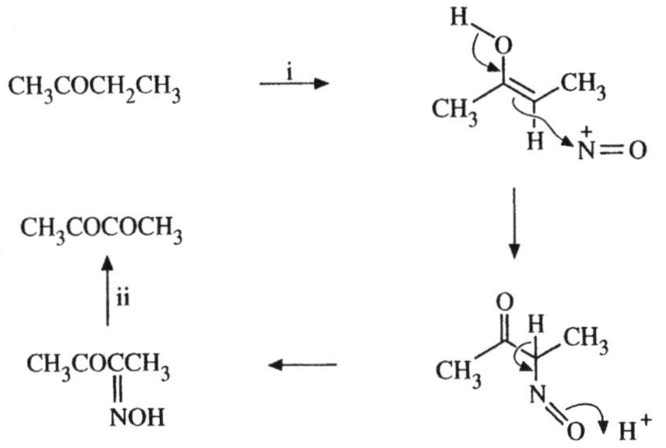

Conditions: i, NaNO$_2$/H$_3$O$^+$; ii, H$_3$O$^+$/Δ

Note that in this example functionalization occurs selectively at the α-CH$_2$ group. This is because the more highly substituted of the two possible enols is formed more easily than the other.

Problems

1. At room temperature the equilibrium enol content of ethyl acetoacetate is less than 1% in water, but nearly 50% in hexane.

2. Suggest a route for converting acetic acid into glycine, NH$_2$CH$_2$CO$_2$H.

3. Identify the thermodynamically most acidic hydrogens in CH$_3$COCH$_2$CH$_2$CH$_3$ and CH$_3$COCH$=$CHCH$_3$.

8 Enolate alkylations with alkyl halides

8.1 Introduction

Simple enolates react irreversibly with soft (polarizable) electrophiles such as methyl iodide. A new carbon–carbon bond is formed, in an S_N2 reaction which is of considerable synthetic importance.

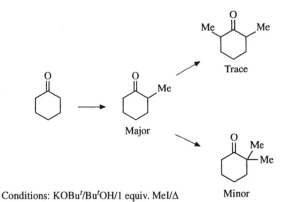

8.2 Enolates derived from simple ketones

If cyclohexanone is treated with ethanolic sodium ethoxide in the presence of methyl iodide, most of the latter reacts with ethoxide ion (Williamson's ether synthesis). However, if the considerably more hindered and slightly more basic *t*-butoxide is used as base, little reaction of it with the methyl iodide occurs. The methyl iodide therefore remains available to react with the ketone enolate as it is generated. From a preparative point of view, however, there is still the complication that the product here has α-CH, so further reaction to give both possible dimethyl products can take place.

Conditions: KOBu^t/Bu^{t}OH/1 equiv. MeI/Δ

The difficulty arises from the fact that substantial concentrations of bases derived from alcohols (pK_a 16–19) are necessary to obtain even small concentrations of enolates from simple ketones (α-CH pK_a about 20). The monoalkyl product is therefore formed in the presence of excess base, and as the reaction proceeds there is increasing competition for the alkylating agent between the monoalkyl product enolate and that of the starting ketone. This problem can be solved by separating the enolization and alkylation stages, which can be done by using one equivalent of a very strong base to (e.g. LDA) to enolize the ketone quantitatively, and then quenching the resulting enolate with a modest excess of alkylating agent.

There is also obviously a regiospecificity problem when it is required to ~alkylate an unsymmetrical ketone such as 2-methylcyclohexanone. A number of ingenious modern methods of achieving control in this situation have been devised. We shall exemplify just one. In 2-methylcyclohexanone, the α-CH$_2$ group loses a proton to strong base more rapidly than the α-CHMe group does. In conventional jargon, the α-CH$_2$ has the greater 'kinetic acidity'. This is partly for steric and statistical reasons, and partly because the electron-releasing effect of the methyl group discourages ionization on its side of the carbonyl group. On the other hand, of the two possible enolates, the one formed by ionization at the α-CHMe group has the lower free energy (i.e. the α-CHMe group has the greater 'thermodynamic acidity'), illustrating the general principle that the more highly alkylated a carbon–carbon double bond is, the more stable it is. Thus 2-methylcyclohexanone reacts with LDA at low temperatures to give the 'kinetic enolate', which can be trapped as its trimethylsilyl ether, but heating with triethylamine and trimethylsilyl chloride gives the trimethylsilyl ether of the 'thermodynamic enolate'.

Thermodynamic enolate   Kinetic enolate

See also Primer 1.

Major product   Major product

Conditions: i, LDA/DME/0 °C; ii, Me$_3$SiCl; iii, Me$_3$SiCl/DMF/Et$_3$N/Δ

Note that, as indicated at the outset, although enolates usually react through their relatively soft and polarizable carbon atoms with soft electrophiles like methyl iodide, they are in fact bifunctional—'ambident'—nucleophiles. Hard electrophiles, and trimethylsilyl chloride is in that class, react at the hard end of the enolate, i.e. oxygen, the site of greatest negative charge density.

The two isomeric trimethylsilyl enol ethers can each be isolated in a pure condition. Their silicon–oxygen bonds can be cleaved by nucleophilic attack at silicon, with methyllithium or a tetraalkylammonium fluoride. This generates in a specific controlled manner, a single enolate which can be caught by addition of an alkylating agent (benzyl bromide in the following examples) before equilibration takes place.

Site of greater negative charge density; hard nucleophile, most reactive to hard electrophiles like Me$_3$SiCl

Site of greater polarizability; soft nucleophile, most reactive to soft electrophiles like MeI

Conditions: i, MeLi/DME/20 °C; ii, PhCH$_2$Br

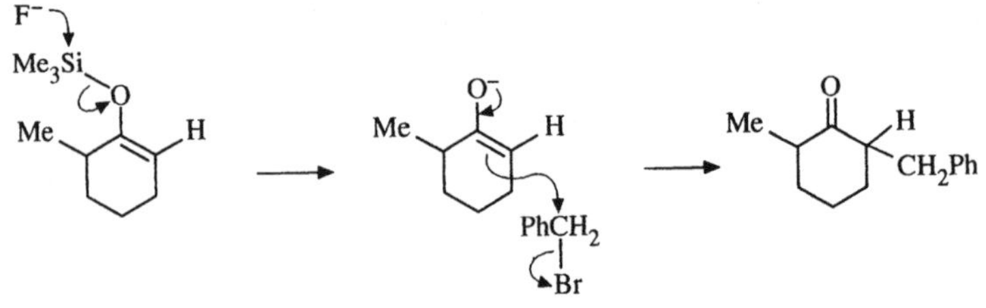

Conditions: $PhCH_2\overset{+}{N}Me_3$. F^- and $PhCH_2Br$ in THF, 0–20 °C

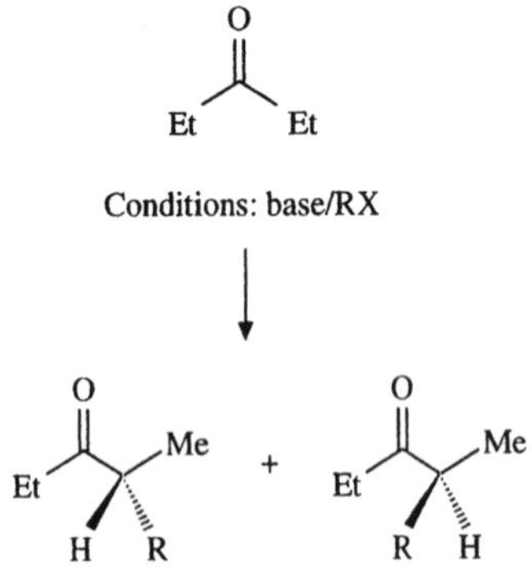

Conditions: base/RX

There is also a stereochemical issue associated with the mono-α-alkylation of simple ketones, except at methyl groups, because a new chiral centre is created. In general, of course, the mono-α-alkyl derivative is produced as a racemic mixture as in the methylation of diethyl ketone. Stereochemical control can, however, be achieved by the temporary derivatization of the carbonyl group with a chiral auxiliary reagent, (*S*)-1-amino-2-methoxylpyrrolidine (SAMP) or its (*R*)-enantiomer (RAMP). The resulting hydrazones α-deprotonate on treatment with LDA, giving chelated anions which react with electrophiles on one side only, so that on removal of the auxiliary group and regeneration of the carbonyl group single enantiomers result (*S* from the SAMP reagent, *R* from the RAMP reagent).

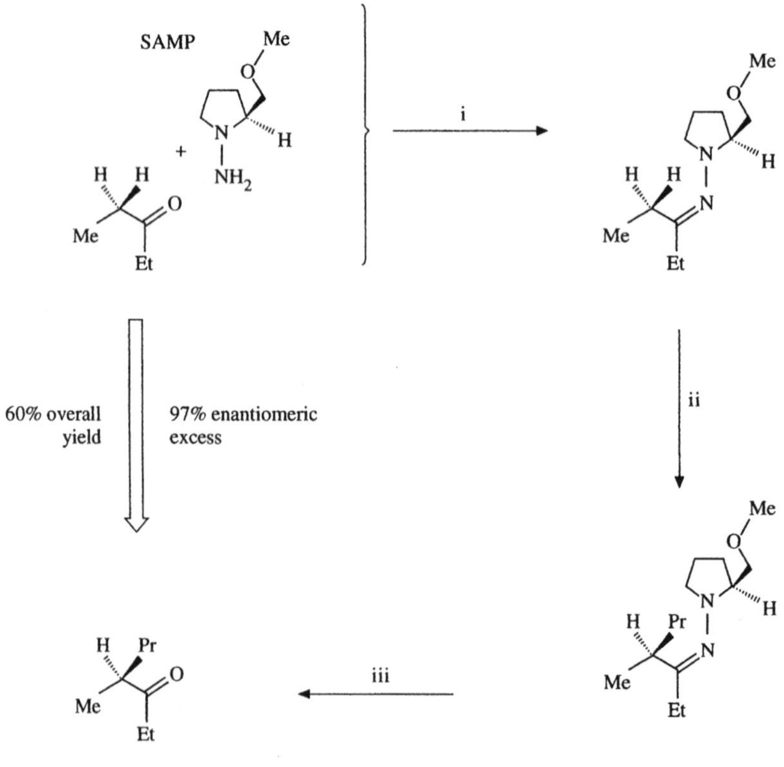

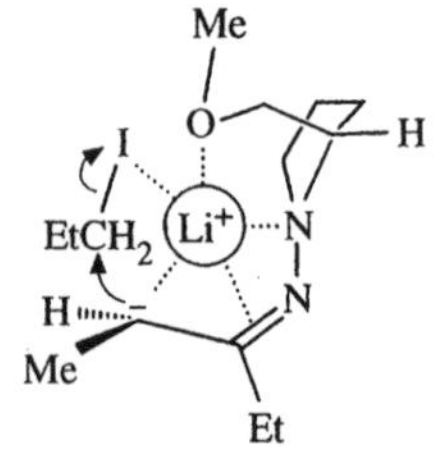

Transition state for stage ii

60% overall yield

97% enantiomeric excess

Conditions: i, 60°C/overnight; ii, LDA/Et₂O/0 °C then –110 °C/PrI; iii, O₃/–78 °C/CH₂Cl₂

8.3   Enolates derived from simple esters

Because the acidity—both kinetic and thermodynamic—of the α-CH in an ester such as ethyl acetate is lower than that of the α-CH in a simple ketone, very strong bases are indispensable for enolate generation. With such bases alkylation presents no difficulty, as in the following example.

$$\text{EtCH}_2\text{CO}_2\text{Me} \xrightarrow[\text{ii}]{\text{i}} \text{Et}_2\text{CHCO}_2\text{Me}$$

Conditions: i, LDA/–78 °C/THF;
ii, EtI/THF–HMPA/–78 °C

8.4   Enolates derived from β-dicarbonyl compounds

When a CH or CH_2 group lies between two carbonyl groups, the enhanced acidity enables complete ionization to be achieved with bases like ethoxide ion, and alkylation with NaOEt/EtOH/RHal systems is facile. Applications of acetoacetic ester and diethyl malonate as substrates in such reactions have been important for over a century, and remain so.

Applications of acetoacetic ester

Treatment of acetoacetic ester with an alkyl halide in the presence of an equivalent amount of sodium ethoxide results in alkylation at the carbon between the carbonyl groups, as exemplified below.

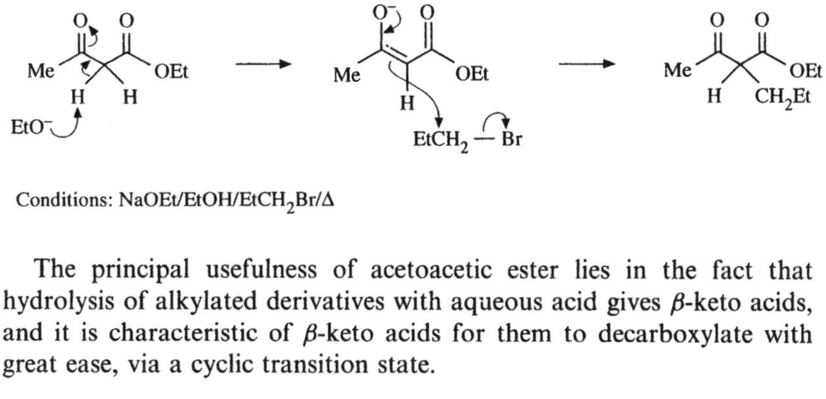

Conditions: NaOEt/EtOH/EtCH$_2$Br/Δ

The principal usefulness of acetoacetic ester lies in the fact that hydrolysis of alkylated derivatives with aqueous acid gives β-keto acids, and it is characteristic of β-keto acids for them to decarboxylate with great ease, via a cyclic transition state.

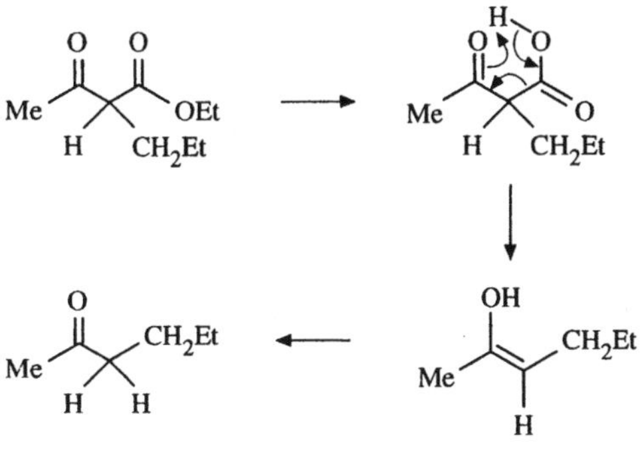

Conditions: dil. NaOH, then dil. H$_2$SO$_4$/Δ

The overall result of an alkylation–hydrolysis–decarboxylation sequence starting with an alkylating reagent RHal is its conversion to RCH_2COCH_3. Since monoalkyl derivatives of acetoacetic ester still have an acidic proton, a second alkylation with a different alkylating reagent is possible under slightly more vigorous conditions, leading by the same work-up operations to a ketone of general structure $RR'CHCOCH_3$.

$$MeCOCH_2CO_2Et \xrightarrow{i} \underset{\underset{R}{|}}{MeCOCHCO_2Et} \xrightarrow{ii} \underset{\underset{R'}{|}}{\overset{\overset{R}{|}}{MeCOCCO_2Et}} \xrightarrow{iii} MeCOCH\overset{R}{\underset{R'}{\diagdown}}$$

Conditions: i, NaOEt/EtOH/RHal; ii, similarly/R′Hal; iii, hydrolysis and decarboxylation

Similar synthetically important transformations can be carried out with other β-keto esters, including cyclic cases. There are two kinds of side-reaction to contend with. Firstly, *O*-alkylation of the enolate may compete significantly with *C*-alkylation. Dipolar aprotic solvents (DMF, DMSO, HMPA) are in many ways suitable, but tend to encourage *O*-alkylation compared to protic solvents, because in the latter the oxygen of the enolate is encumbered by H-bonded solvent molecules. In a series of alkylating agents RCl, RBr, RI, the iodide is the most polarizable and therefore the softest electrophile, so it leads to the least *O*-alkylation, because oxygen is the harder end of the enolate. The iodide is therefore the reagent of choice, subject to economy and availability. Secondly, alkaline hydrolysis of β-keto esters is usually accompanied by some carbon–carbon bond cleavage, leading to two carboxylic acids on acidic work-up.

$$\underset{HO^-}{\underset{Me}{\diagup}}\overset{O \quad O}{C\diagdown R \; R'}OEt \longrightarrow MeCO_2H \; + \; \underset{R' \quad OEt}{\overset{R \quad O^-}{C=C}} \longrightarrow RR'CHCO_2H$$

Hydrolysis under acidic conditions may therefore be preferred, provided it is compatible with other aspects of the case in hand.

Applications of diethyl malonate

Diethyl malonate can be alkylated under similar conditions to those used with acetoacetic ester, in one or two stages.

Hydrolysis with aqueous acid gives a free malonic acid, which decarboxylates once by a mechanism analogous to the decarboxylation of β-keto acids. Diethyl malonate—often simply called malonic ester in casual laboratory talk—thus gives easy access to substituted acetic acids

$$CH_2(CO_2Et)_2 \xrightarrow{i} RCH(CO_2Et)_2 \xrightarrow{ii} RR'C(CO_2Et)_2$$

Diethyl malonate

$$\Big\downarrow iii$$

$$RR'CHCO_2H$$

Conditions: i, RHal/NaOEt/EtOH;
ii, R′Hal, similarly; iii, H_3O^+/Δ

of general formula $RR'CHCO_2H$: the preparation of cyclobutane carboxylic acid is a particularly interesting example.

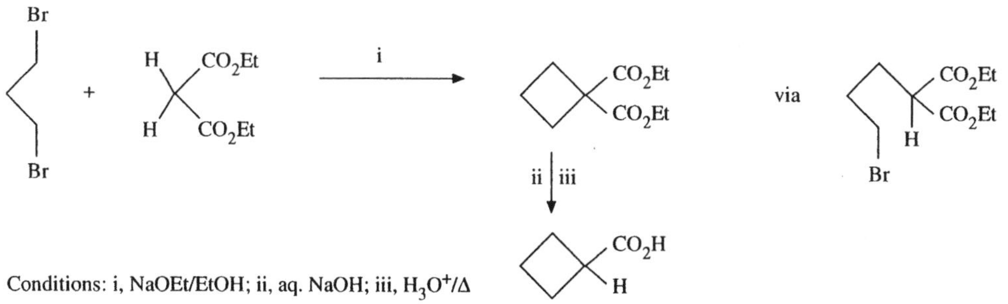

Conditions: i, NaOEt/EtOH; ii, aq. NaOH; iii, H_3O^+/Δ

There are a number of other important applications, of which we shall note just one, the use of diethyl acetamidomalonate.

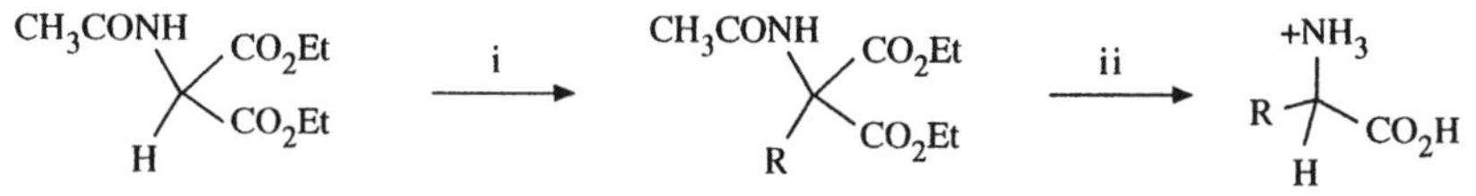

Diethyl acetamidomalonate

Conditions: i, RX/NaOEt/EtOH; ii, H_3O^+/Δ

The acetamido substituent does not interfere with enolate formation, and α-alkylation proceeds smoothly in the presence of sodium ethoxide. Hot aqueous acid hydrolyses not only the two ester groups but also the amide, and decarboxylation takes place under these conditions, giving an α-amino acid. This is one of the most general syntheses of the proteinogenic α-amino acids and their analogues.

Dianions

Acetylacetone enolizes on treatment with alkoxide bases exclusively by loss of a single proton from the CH_2 group, which has vastly greater kinetic and thermodynamic acidity than the methyl groups. Electrophiles therefore react at the CH_2 group in the presence of alkoxide bases. However, if two equivalents of a very strong base are used, a dianion is formed, the second proton being taken from a methyl group. Now if one equivalent of an alkylating agent is added, it is attacked by a former methyl carbon.

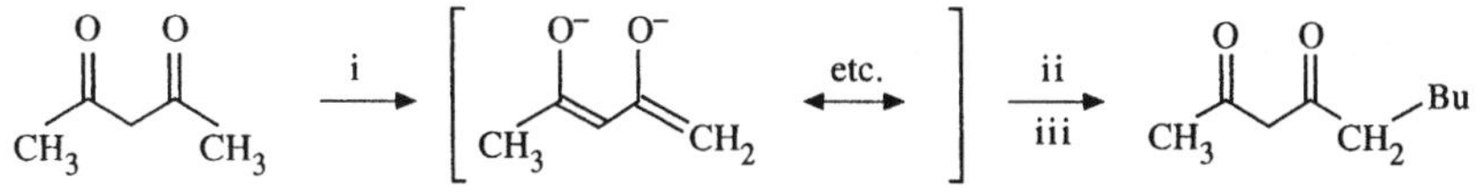

Conditions: i, 2 equiv. KNH_2/liq. NH_3; ii, 1 equiv. BuBr; iii, H_3O^+

Analogous procedures can be carried out on other β-dicarbonyl and related compounds: in unsymmetrical cases, where two dianions could in principle be formed, in practice one of the two is often formed more easily than the other.

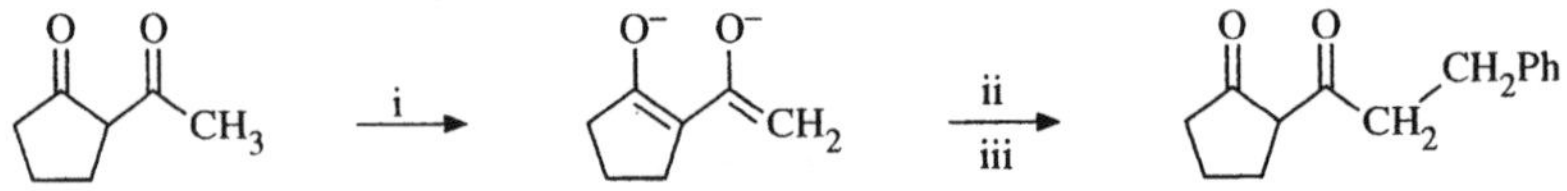

Conditions: i, 2 equiv. KNH$_2$/liq. NH$_3$; ii, PhCH$_2$Cl; iii, H$_3$O$^+$

Both of the last two examples illustrate the general rule that in dianions with two nucleophilic carbons, it is the one corresponding to the less acidic site of the starting material which reacts fastest with electrophiles. The same rule applies to dianions derived from carboxylic acids, which are also useful synthetic intermediates.

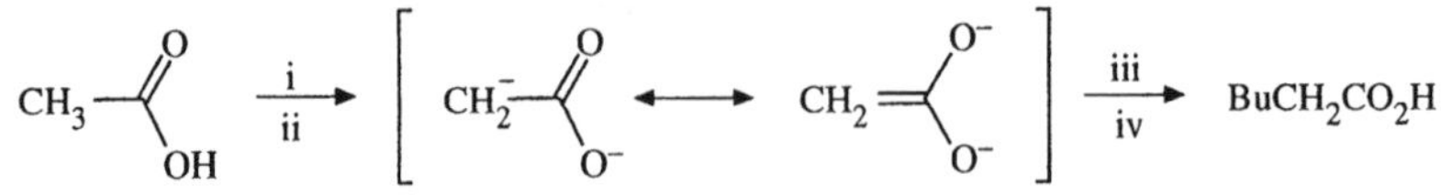

Conditions: i, 2 equiv. LDA/THF/–20 °C; ii, HMPA/50 °C/2 h; iii, BuBr; iv, H$_3$O$^+$

Problems

Explain the following.

1.

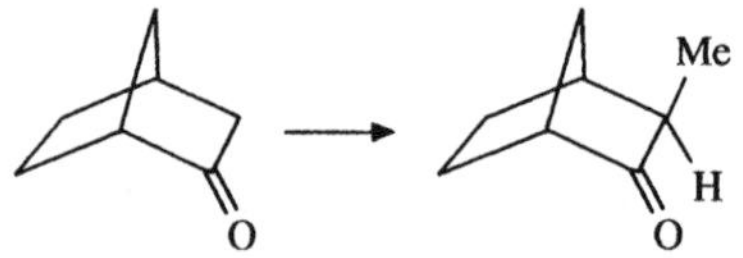

Almost sole product

Conditions: i, Ph$_3$CNa/Et$_2$O; ii, MeI; iii, mild H$_3$O$^+$ work-up

2.

CH$_2$(CO$_2$Et)$_2$ $\longrightarrow$ Butyrolactone

Conditions: i, NaOEt/EtOH; ii, ethylene oxide; iii, Δ/H$_3$O$^+$

3.

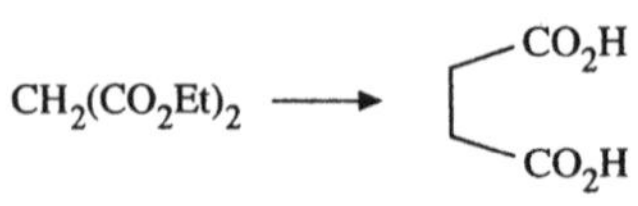

CH$_2$(CO$_2$Et)$_2$ $\longrightarrow$ Succinic acid

Conditions: i, NaOEt/EtOH; ii, I$_2$; iii, Δ/H$_3$O$^+$

9 Aldol condensations and related reactions

9.1 Introduction

The reactions of enols and enolates with electrophiles are not confined to the simple α-substitutions so far discussed. The electrophile can also be a carbonyl compound, and, as with the attack of simpler nucleophiles on carbonyl groups, the formation of a tetrahedral adduct can be followed by protonation, dehydration, or loss of a leaving group.

9.2 The aldol condensations of simple aldehydes and ketones

Acetaldehyde forms a dimer reversibly under the influence of base.

$$2CH_3CHO \; \rightleftharpoons \; \underset{\text{Aldol}}{CH_3\overset{\displaystyle OH}{\overset{|}{C}}HCH_2CHO}$$

The dimer in this specific case has the trivial name aldol (*ald*ehyde–alcoh*ol*), and this is also the name given to the whole class of β-hydroxycarbonyl compounds which can be formed in this way. The reaction is called the aldol condensation, which is rather misleading, as the classical meaning of the term 'condensation reaction' is a reaction in which two molecules combine with loss of water. The mechanism of the base-catalysed formation of aldol itself is as shown below.

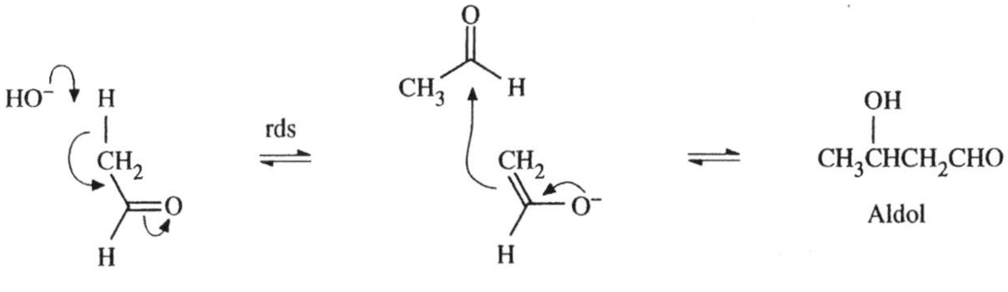

The equilibria favour aldol. At all except very low concentrations, the rate of aldol formation is proportional to $[MeCHO][OH^-]$, which is consistent with enolate formation being the rate-determining step. Treatment of acetone with aqueous base leads to corresponding equilibria being established, but in this instance the product, diacetone alcohol, is not favoured, and the rate of its formation depends on $[Me_2CO]^2[OH^-]$, which is consistent with carbon–carbon bond formation being rate-determining in this case.

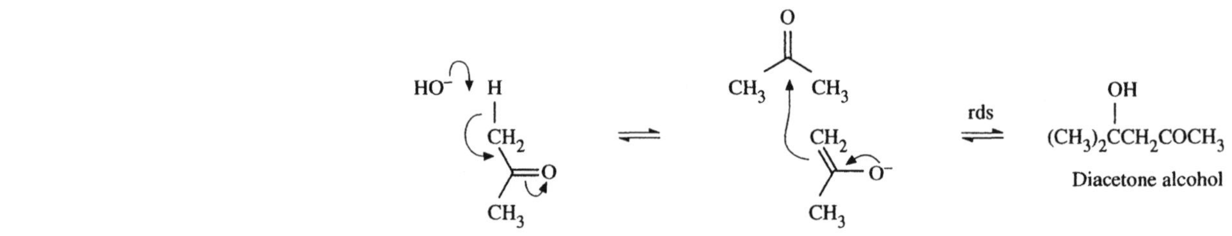

In order to use the unfavourable equilibria shown to prepare diacetone alcohol from acetone, a clever practical trick is employed. Acetone is refluxed in such a way that *en route* back to the flask the distillate percolates through some $Ba(OH)_2$, an insoluble basic catalyst. Equilibration takes place while there is contact with the catalyst, and it is an equilibrium mixture which returns to the flask. As acetone has a lower boiling point than diacetone alcohol, it is pure acetone which distills up to be passed through the catalyst again; the concentration of diacetone alcohol builds up in the flask because there is no base catalyst there to make it revert to acetone.

These aldol condensations are reversible because the carbon–carbon bond formed in the reaction can break under the influence of base by a cleavage mechanism which is general for β-hydroxycarbonyl compounds—the reverse or retro aldol reaction.

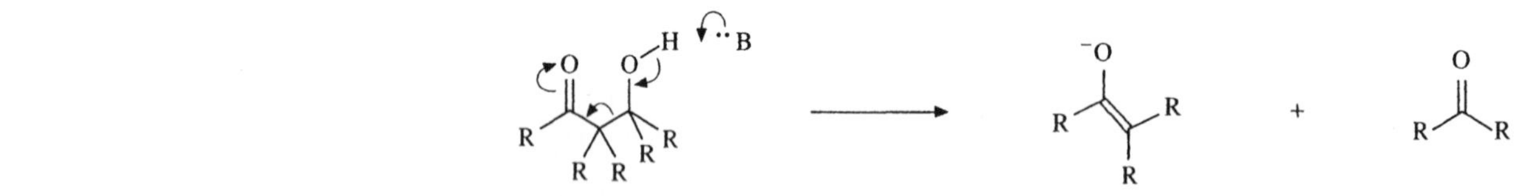

Acidic catalysis also induces aldol condensations; again, they are reversible, but they are complicated by scope for further reactions.

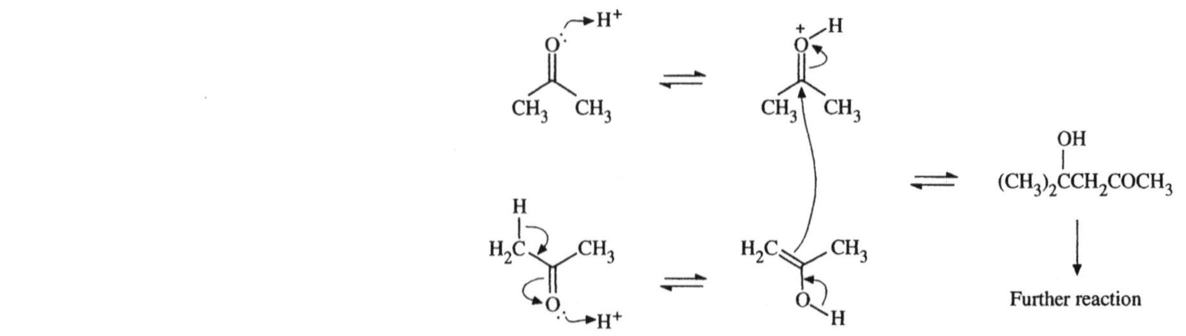

Aldol condensations, whether carried out with acid or base catalysis, are often followed by spontaneous dehydration.

Dehydration, which is driven by the conjugation present in the resulting $\alpha\beta$-unsaturated system, is practically the norm for acid-catalysed conditions.

9.3 Crossed aldol condensations

The aldol condensations which were outlined in Section 9.2 are special very simple cases, in which there is only one possible β-hydroxycarbonyl product. With unsymmetrical ketones, there is regiochemical and stereochemical ambiguity: three racemic products can in principle result, as in the base-catalysed aldol condensation of methyl ethyl ketone.

Base-catalysed aldol condensation of methyl ethyl ketone—only one enantiomer of each possible product is shown.

The situation can become quite imponderable if a crossed condensation, i.e. a reaction in which the electrophile and nucleophile in the product-forming step come from different starting materials, is attempted simply by mixing the two components under basic conditions. Regiochemical and stereochemical control can be achieved by using specifically preformed lithium or boron enolates, but these elegant procedures are beyond the scope of this Primer. However, crossed all-in-together aldol reactions are unambiguous and useful if certain conditions are satisfied. If one reactant is more electrophilic than the other, and that more electrophilic reactant lacks α-CH, then a single product results. Formaldehyde thus reacts unambiguously with 2-methylpropionaldehyde.

$$Me_2CHCHO \longrightarrow Me_2CCHO$$

Conditions: CH₂O/dil. aq. Na₂CO₃

Indeed, if formaldehyde is present in excess under basic conditions, it will replace every α-CH of any other aldehyde or ketone.

Conditions: CH_2O/aq. $Ca(OH)_2$

With aldehydes and excess formaldehyde in strongly basic conditions, a Cannizzaro reaction (see Section 2.8) follows replacement of all the α-CH.

$$CH_3CHO \longrightarrow C(CH_2OH)_4 \quad \text{via} \quad HOCH_2-\overset{\overset{\displaystyle CH_2OH}{|}}{\underset{\underset{\displaystyle CH_2OH}{|}}{C}}-CHO$$

Conditions: xs CH_2O/conc. KOH/Δ

Crossed reactions between ketones with α-CH_2 and benzaldehyde also take place readily, usually followed by spontaneous dehydration, because this gives an extended conjugated system.

$$PhCHO + CH_3COPh \longrightarrow PhCH=CHCOPh$$

Conditions: NaOH/aq. EtOH/Δ

The ability to form so-called benzylidene or benzal derivatives was classically a criterion for the presence of an 'activated' CH_2 group. The reaction between acetone and excess benzaldehyde takes place on both sides of the ketone carbonyl.

$$xs\ PhCHO + CH_3COCH_3 \longrightarrow PhCH=CHCOCH=CHPh$$
Dibenzylidene acetone

Conditions: NaOH/aq. EtOH/Δ

Ethyl acetate also reacts with benzaldehyde, but aqueous sodium hydroxide is not basic enough, and is inappropriate because it would cause ester hydrolysis: sodium ethoxide in ethanol is therefore used.

$$PhCHO + CH_3CO_2Et \longrightarrow PhCH=CHCO_2Et$$
Ethyl cinnamate

Conditions: NaOEt/EtOH/0 °C

The Knoevenagel condensation of malonic acid and its derivatives is of the same type, but weaker base catalysis is sufficient because of the ease with which malonic acid and its derivatives enolize.

$$PhCHO + CH_2(CO_2Et)_2 \longrightarrow PhCH=C(CO_2Et)_2$$

Conditions: pyridine/Δ

If malonic acid itself is used, decarboxylation takes place after the aldol reaction, by an elimination mechanism.

PhCHO + CH$_2$(CO$_2$Et)$_2$ ⟶ PhCH=CHCO$_2$H

Trans-(E)-cinnamic acid

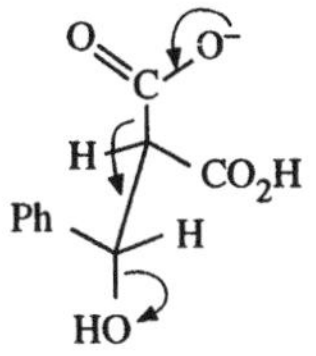

Conditions: pyridine/cat. amt. piperidine/Δ; H$_3$O$^+$ work-up

via

In the Perkin reaction, acetate ion is the base and acetic anhydride is the source of the enolate reactant.

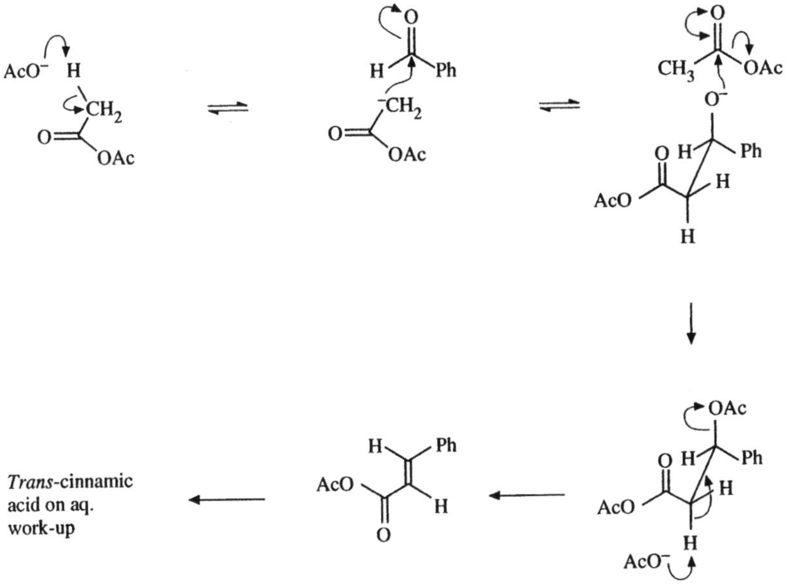

Conditions: PhCHO/Ac$_2$O/KOAc/Δ

Trans-cinnamic acid on aq. work-up

Note that in the Perkin reaction, the alkoxide oxygen of the tetrahedral adduct is trapped by acetylation, which opens the way to the effectively irreversible elimination of acetic acid. The Stobbe condensation has a similar critical feature.

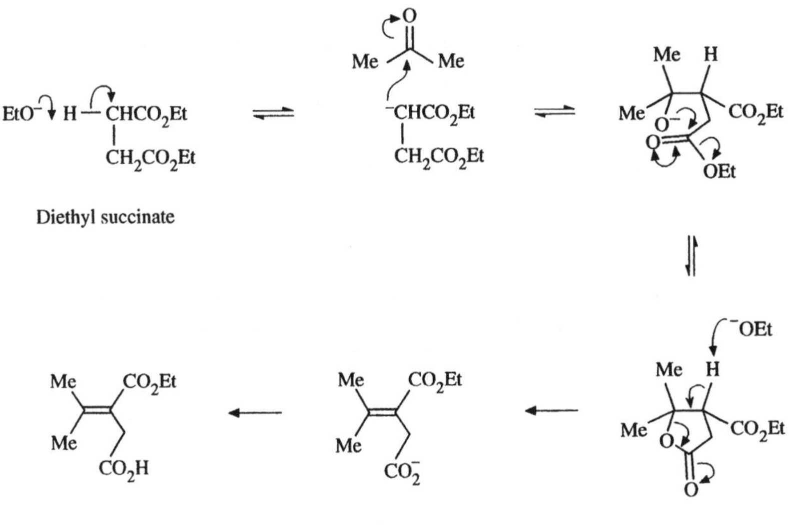

Diethyl succinate

Conditions: NaOEt/Et$_2$O/−15 to 23 °C/ several days; alternatively KOBu^t/Bu^{t}OH/Δ/0.5 h; H$_3$O$^+$ work-up

Here the alkoxide oxygen of the initial adduct is caught by lactonization, which ultimately results in formation of a carboxylate anion. In the prevailing basic conditions, the equilibria which lead to this carboxylate anion are all displaced in its favour, and none of the other reversible pathways which could in principle be followed can compete. Under the conditions of the reaction, acetone could self-condense (see above), and so could diethyl succinate (see the next section). A crossed condensation the other way round—i.e. attack of acetone enolate on an ethoxycarbonyl group—would be no surprise, because that is in fact what happens if acetone and ethyl acetate are treated with sodium ethoxide in ethanol.

$$CH_3COCH_3 + CH_3CO_2Et \longrightarrow CH_3COCH_2COCH_3$$

Conditions: NaOEt/EtOH/Δ; H_3O^+ work-up

None of these possibilities, however, give an end product which has acidity anything like as great as a carboxylic acid, so none of them is favoured by displacement of the relevant end-of-the-line equilibrium. The Stobbe reaction is subject to thermodynamic control. The Darzens reaction also involves the trapping of the alkoxide oxygen of the tetrahedral adduct, and only one product is isolated despite the potential of the reactants to lead to others.

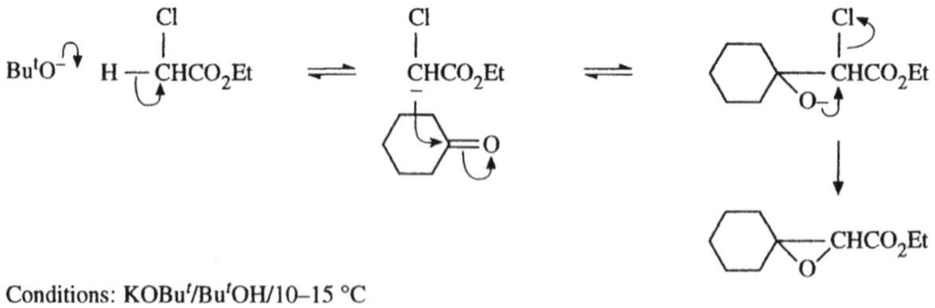

Conditions: KOBu^t/Bu^{t}OH/10–15 °C

In this case kinetic control is operative. The epoxide-ring closure is effectively irreversible, because chloride ion is a poor nucleophile versus such substrates. The other possible products, in contrast, are related to the starting materials by freely reversible equilibria. This situation has the consequence that eventually all the reactants are converted to the epoxide.

The crossed aldol condensations surveyed above all work because the structures of the reactants in some way dictate that one of several possible products is favoured. What of the case where it is desired to go against Nature? Consider the potential crossed aldol reaction shown below.

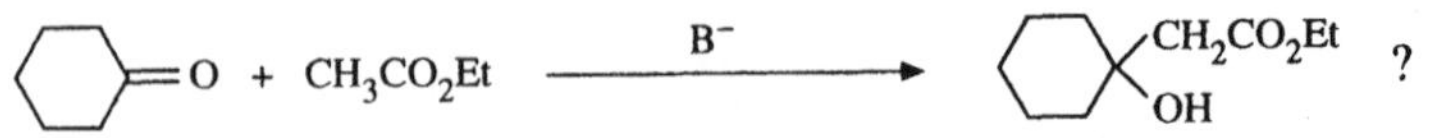

As already mentioned, attempting this sort of condensation as shown fails completely, as different pathways leading to β-diketones are generally preferred. However, the required conversion can be achieved by a Reformatsky reaction.

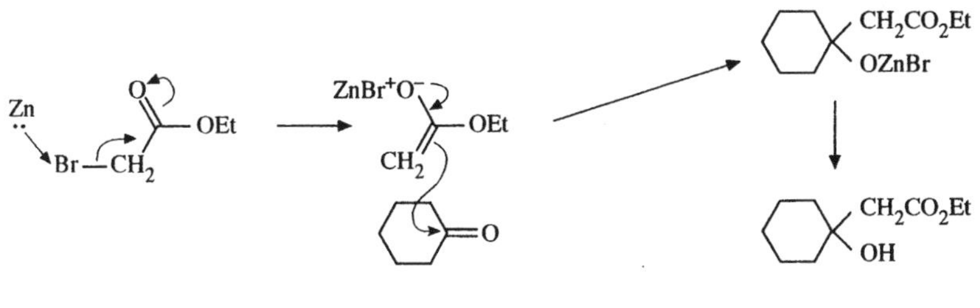

Conditions: bromoester/cyclohexanone/Zn all in together/PhH/100 °C; work-up after cautious acidification

Why must the acidification be done under mild conditions?

This procedure works for two reasons. Firstly no strong base is present, so none of the possible base-catalysed condensations take place. Secondly the zinc enolate intermediate which is formed is stabilized, relatively unreactive, and therefore selective, so it reacts preferentially with the more reactive ketone carbonyl group. A modern way of procuring reaction between an ester and a ketone with the same outcome is to generate the enolate of the ester quantitatively using a lithium amide base, and then quench it with the ketone, as in the eqn 9.21.

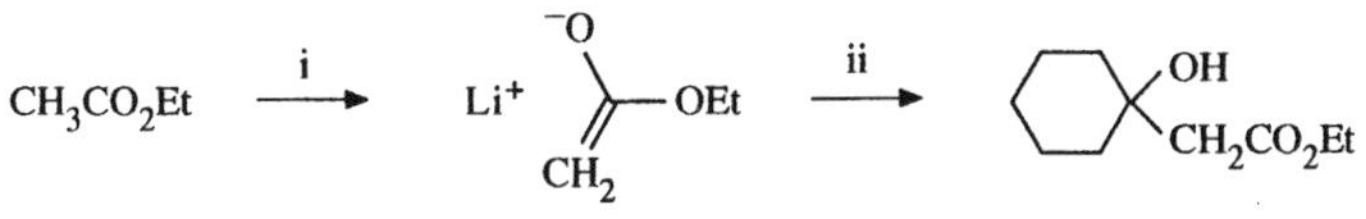

Conditions: i, LiN(SiMe$_3$)$_2$/THF; ii, cyclohexanone; cautious H$_3$O$^+$ work-up

Acid-catalysed crossed aldol condensations are often complex in outcome, and are only useful in special instances. However, there is an analogue in the Mannich reaction which is of synthetic importance. It is exemplified by the reaction of acetone, formaldehyde, and diethylamine under mildly acidic conditions.

$$CH_3COCH_3 + CH_2O + Et_2\overset{+}{N}H_2.Cl^- \longrightarrow CH_3COCH_2CH_2NEt_2$$

Conditions: Me$_2$CO/paraformaldehyde/Et$_2\overset{+}{N}$H$_2$.Cl$^-$/cat. amt. HCl/Δ, then basification before work-up

Here the electrophile, which is formed by an addition–dehydration reaction between formaldehyde and diethylamine,

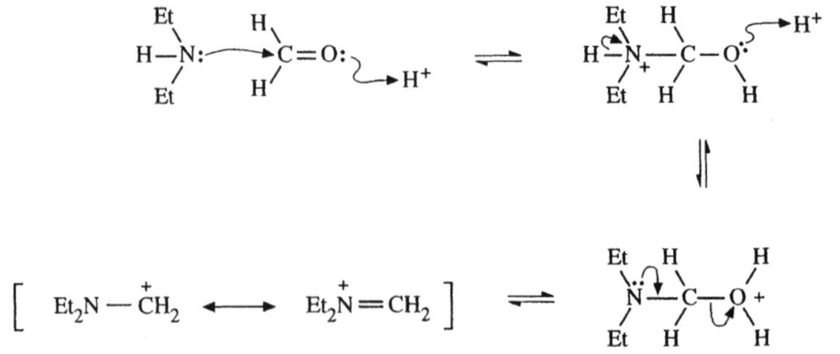

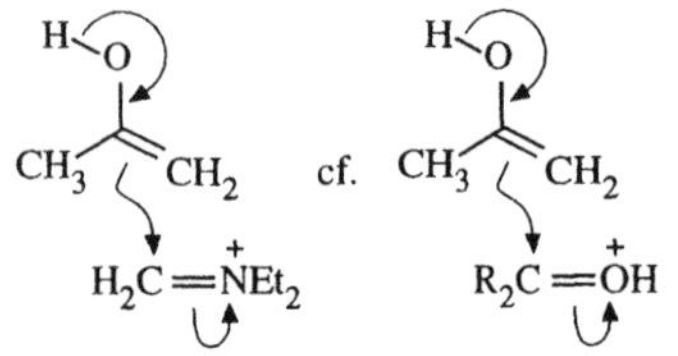

$$\left[\; Et_2N - \overset{+}{C}H_2 \; \longleftrightarrow \; Et_2\overset{+}{N}=CH_2 \; \right] \rightleftharpoons$$

is analogous to a protonated carbonyl group, and reacts with acetone enol by analogy with an acid-catalysed aldol condensation.

The products of Mannich reactions—called Mannich bases—have many synthetic applications, and Mannich-type reactions are also involved in the biosynthesis of several important groups of alkaloids.

See Primer 20.

9.4   Enolate acylation reactions

Enolates attack acylating agents.

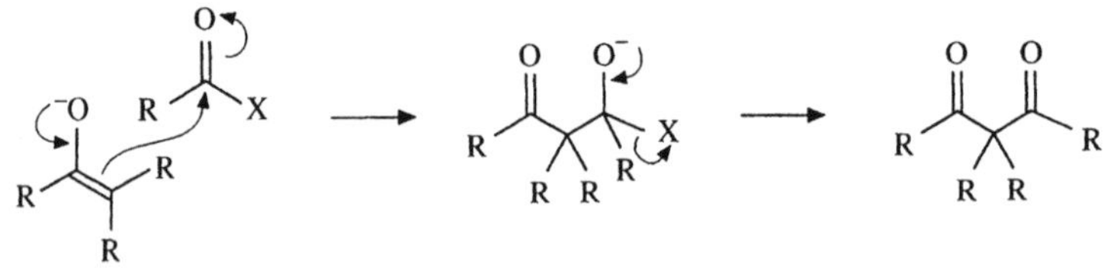

This is a very general reaction; it is reversible. The case X = OMe or OEt is the most important. The simplest example is the Claisen ester self-condensation of ethyl acetate.

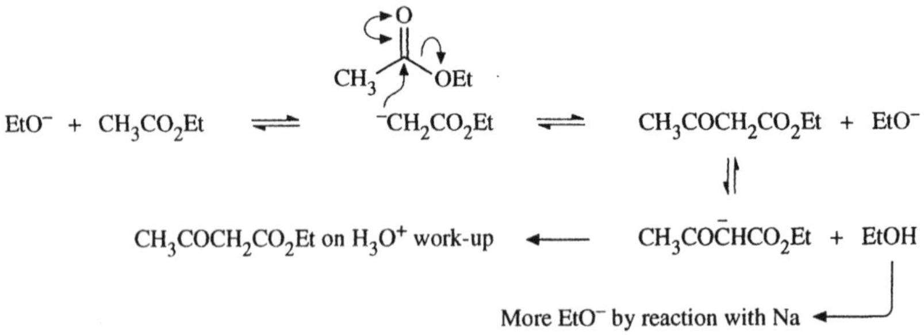

$$EtO^- + CH_3CO_2Et \;\rightleftharpoons\; {}^-CH_2CO_2Et \;\rightleftharpoons\; CH_3COCH_2CO_2Et + EtO^-$$

$CH_3COCH_2CO_2Et$ on H_3O^+ work-up $\longleftarrow$ $CH_3CO\overset{-}{C}HCO_2Et$ + EtOH

More EtO⁻ by reaction with Na $\longleftarrow$

Conditions: Na/CH_3CO_2Et/cat.amt. EtOH/Δ

A full equivalent of sodium ethoxide base is needed here, although this is actually conveniently supplied by starting with a full equivalent of metallic sodium and a small amount of ethanol, which react to give sodium ethoxide: as the reaction proceeds, ethanol is produced, and thence more sodium ethoxide. The full equivalent of base is necessary to ionize the product, in order to pull over all the equilibria in its favour. If the β-keto ester which would be formed has no acidic α-CH, then the reaction fails if attempted with sodium ethoxide. In such a case, however, the use of a much stronger base such as $Ph_3C^-Na^+$ drives the reaction by generating the enolate of the starting material quantitatively.

$$Me_2CHCO_2Et \;\;\times\!\!\!\to\;\; Me_2CHCOCCO_2Et$$
$$\overset{Me}{\underset{Me}{|}}$$

but

$$Me_2CHCO_2Et \;\longrightarrow\; Me_2CHCOCCO_2Et$$
$$\overset{Me}{\underset{Me}{|}}$$

Conditions: NaOEt/Δ; H_3O^+ workup

Conditions: $Ph_3CNa/Et_2O/25$ °C; H_3O^+ work-up

As with crossed all-in-together aldol condensations, crossed all-in-together Claisen ester condensations work only if one reactant is both more electrophilic than the other and also lacks α-CH.

$$CH_3CH_2CO_2Et \;+\; \overset{CO_2Et}{\underset{CO_2Et}{|}} \;\longrightarrow\; \overset{CH_3CHCO_2Et}{\underset{COCO_2Et}{|}}$$

Conditions: NaOEt/Et_2O; H_3O^+ work-up

A more general approach to ester enolate acylation is to separate the enolization and acylation stages.

$$CH_3CH_2CO_2Et \;\xrightarrow[\text{ii}]{\text{i}}\; CH_3\overset{COCH_2CH_3}{\underset{CO_2Et}{\overset{|}{CH}}}$$

Conditions: LDA/THF/−78 °C; CH_3CH_2COCl/−78 °C; H_3O^+ work-up

An important variant of the Claisen ester condensation, which is in all fundamentals exactly like it but nevertheless has its own name—the Dieckmann reaction—is where the reacting groups belong to the same molecule, resulting in the formation of a carbocyclic ring

Diethyl adipate

Conditions: Na/PhH/cat.amt.
EtOH/Δ; H_3O^+ work-up

Good yields of six-membered ring β-keto esters are also obtained from the appropriate diester. The reaction is less satisfactory for large rings because the probability of productive encounter between the reacting groups falls off as the chain length between them increases. With rings of intermediate size (say 8–12 carbons in the ring), there are unfavourable steric interactions between ring substituents as well. The Dieckmann reaction also fails with strained small-ring targets. Diethyl succinate, for example, undergoes intermolecular Claisen ester condensation and then a Dieckmann-type cyclization, to give a six-membered ring.

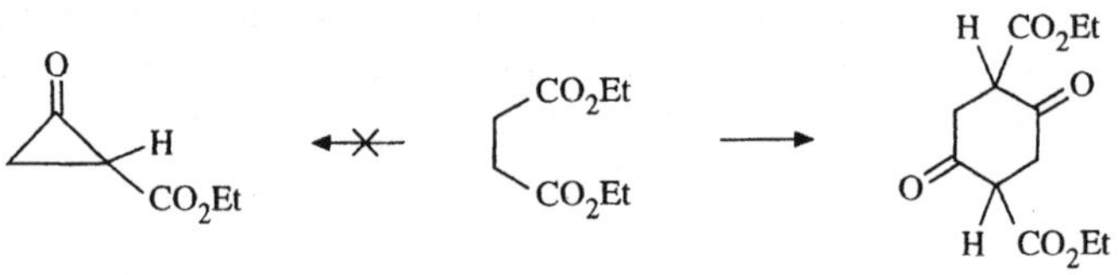

Conditions: NaOEt/Δ; H_3O^+ work-up

Ketone enolates are also reversibly acylated by carboxylic acid esters, as already exemplified by the reaction between acetone and ethyl acetate. If an alkoxide base is used under equilibrating conditions, a full equivalent of base is required. The strongest acid among the possible products, acetylacetone, is formed, because the various equilibria are all displaced in its favour by the base. The very fact that makes a single

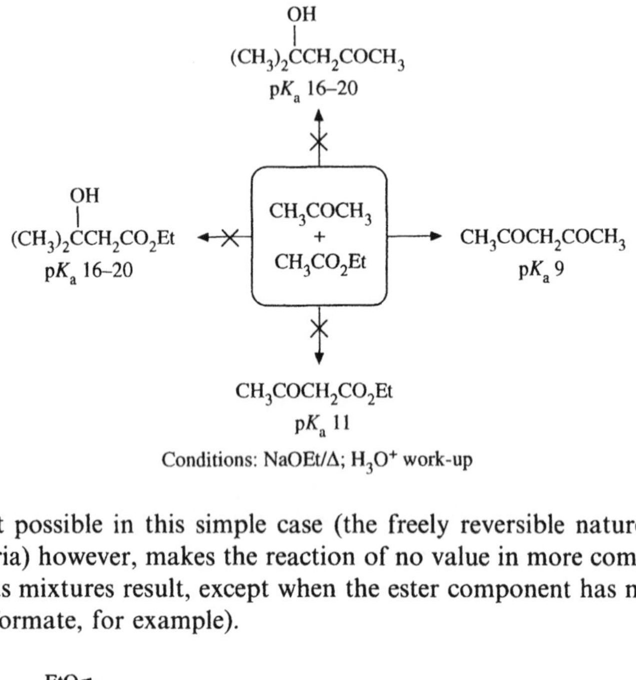

Conditions: NaOEt/Δ; H_3O^+ work-up

product possible in this simple case (the freely reversible nature of the equilibria) however, makes the reaction of no value in more complicated cases, as mixtures result, except when the ester component has no α-CH (ethyl formate, for example).

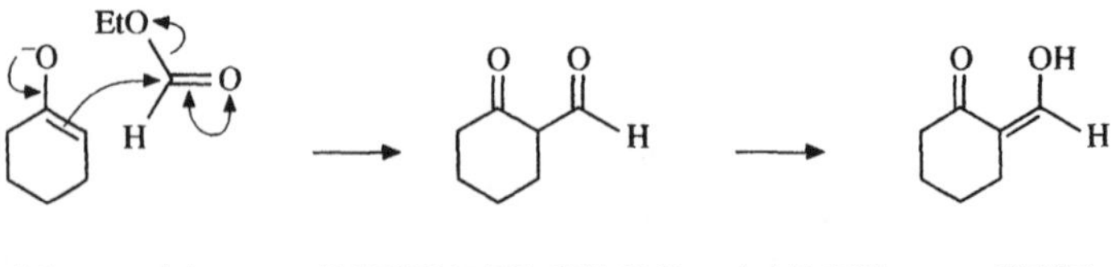

Conditions: cyclohexanone/EtOCHO(xs)/Et_2O/NaH (1 equiv.)/EtOH(cat. amt.)/20 °C; the product is almost entirely in the enolic form after H_3O^+ work-up

One of Nature's ways of making new carbon–carbon bonds uses a thiol ester analogue of the Claisen ester condensation, taking advantage of some key facts about thiol esters. Firstly, they are more reactive to nucleophilic attack at the carbonyl group than ordinary esters. Secondly, their α-CH acidity is greater. These differences arise because in an ordinary ester electron-release by the ester oxygen to the carbonyl carbon diminishes (compared to the situation in a ketone) the electrophilic character of that carbon, and destabilizes the corresponding enolate.

This conjugation arises from overlap of the sp^2-hybridized ester oxygen with the carbonyl π-system. In a thiol ester, the orbitals of the sulfur atom are larger and more diffuse, so they do not overlap effectively with the carbonyl π-system. Despite the fact that the electronegativity of sulfur is less than that of oxygen, there is less electron release by sulfur. To this extent, thiol esters are more 'ketone-like' than ordinary esters. This is not, however, a helpful analogy, because thiol esters do not undergo typical ketone reactions like oxime formation: RS^- is in fact a better leaving group than RO^-, so nucleophilic addition to the carbonyl group invariably leads to overall substitution, not condensation.

The sulfur atoms involved in biological Claisen condensations and related reactions are attached to complex carrier molecules or enzymes, but this does not affect the essential chemistry, which is shown below.

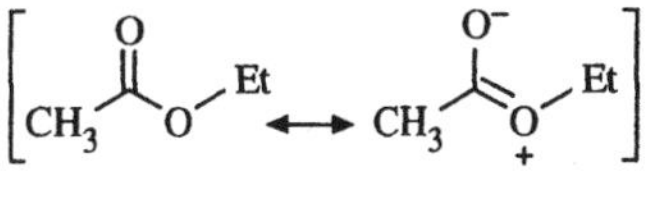

This canonical diminishes carbonyl electrophilicity

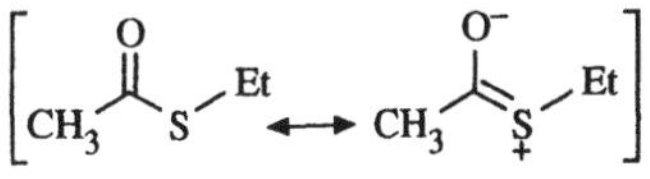

This canonical contributes less than in MeCO₂Et

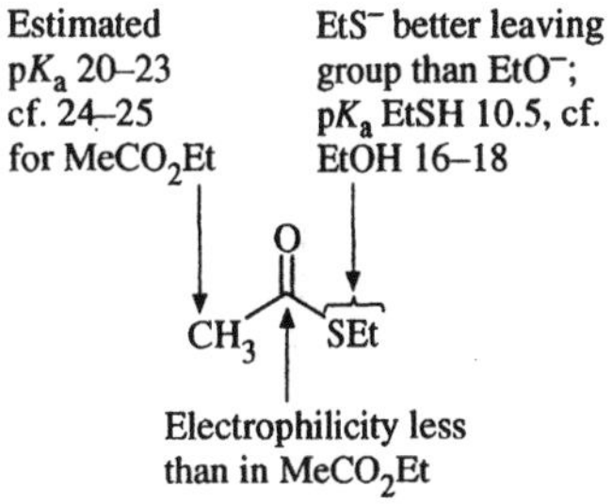

Estimated pKₐ 20–23 cf. 24–25 for MeCO₂Et

EtS⁻ better leaving group than EtO⁻; pKₐ EtSH 10.5, cf. EtOH 16–18

Electrophilicity less than in MeCO₂Et

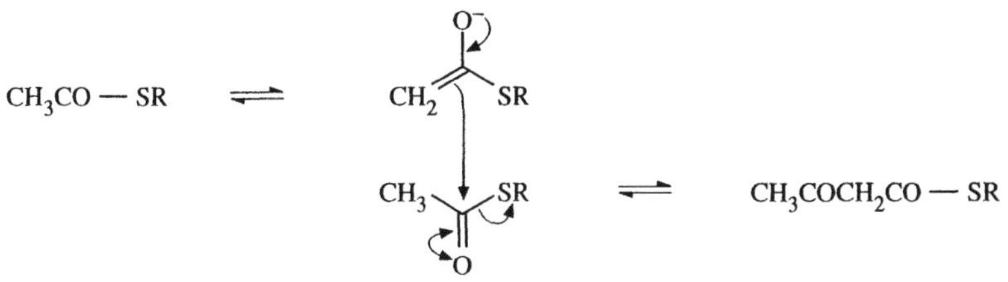

9.5 The Thorpe–Ziegler cyclization: nitrile groups as carbonyl analogues

Much of the chemistry of organic cyanides—nitriles, RCN—is analogous to that of carboxylic esters. Firstly, strong bases cause α-deprotonation to give an anion which is stabilized by delocalization as in an enolate: pK_a CH₃CN 25, cf. pK_a CH₃CO₂Et 25. Such an anion has a soft nucleophilic α-carbon like an enolate.

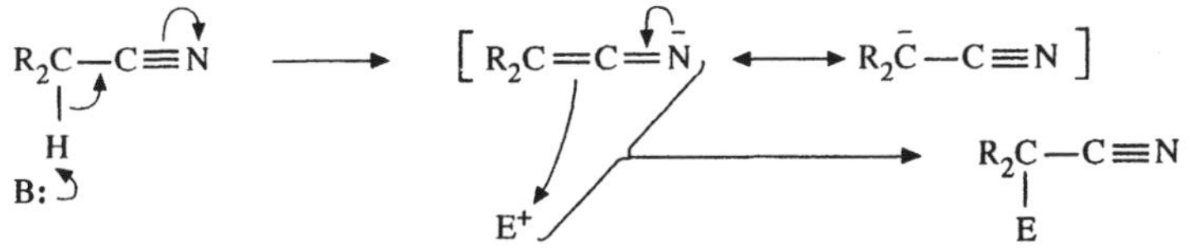

Secondly, the nitrile carbon is electrophilic, and reactive nucleophilic reagents NuH can add across the triple bond, giving an imine derivative which is easily hydrolysed on aqueous work-up.

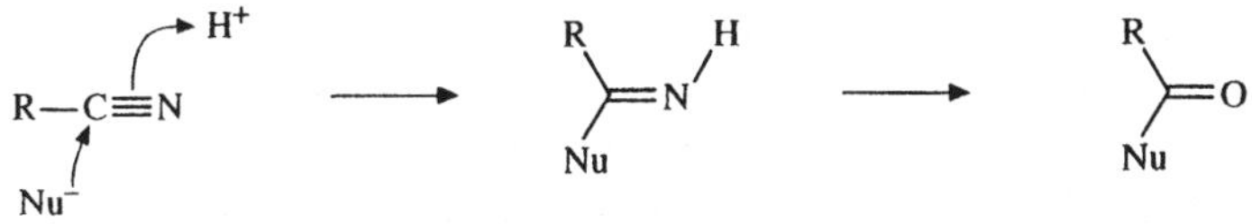

If the nucleophilic reagent is water at the outset, rather vigorous conditions are required, and the amide produced by addition and hydrolysis is further hydrolysed, giving the corresponding carboxylic acid.

$$RC\equiv N \ + \ H_2O \ \longrightarrow \ RCONH_2 \ \longrightarrow \ RCO_2H \ + \ NH_3$$

Conditions: strongly acidic or strongly akaline followed by acidic work-up

All these features are demonstrated in the Thorpe–Ziegler synthesis of cyclic ketones, an example of which is set out below.

Conditions: i, dinitrile added slowly to a mixture containing PhMeN⁻Li⁺ in a large volume of refluxing Et_2O; ii, mineral acid/Δ work-up

9.6 Enolates and $\alpha\beta$-unsaturated carbonyl compounds

Protonation of the adduct at carbon results in overall addition of NuH to the carbon–carbon double bond

We noted in Section 2.1 that in $\alpha\beta$-unsaturated carbonyl compounds the electrophilic reactivity of the carbonyl carbon is diminished by conjugation: the reactivity of the alkene moiety to electrophiles is also depressed, and it becomes receptive to nucleophiles. The $\alpha\beta$-unsaturated system thus behaves like an extended carbonyl group, and nucleophilic reagents like $LiAlH_4$ and Grignards can attack the β-position as well as the carbonyl carbon (see Section 2.6 and 2.9, respectively). Enolates are softer nucleophiles than $LiAlH_4$ and Grignards when reacting through their carbon atoms and, with $\alpha\beta$-unsaturated carbonyl compounds as substrates, do so selectively at the β-position. This happens in the sodium ethoxide-induced reaction of diethyl malonate and ethyl cinnamate, which is an example of the Michael reaction.

$$PhCH=CHCO_2Et \ + \ CH_2(CO_2Et)_2 \ \longrightarrow \ \underset{\underset{CH(CO_2Et)_2}{|}}{PhCH-CH_2CO_2Et}$$

Conditions: NaOEt; H_3O^+ work-up

Such reactions may take place spontaneously after aldol condensation–dehydration sequences, as in the example below.

$$CH_2O \ + \ CH_2(CO_2Et)_2 \ \longrightarrow \ (EtO_2C)_2CHCH_2CH(CO_2Et)_2 \quad via \ \ CH_2=C(CO_2Et)_2$$

Conditions: $CH_2(CO_2Et)_2$/aq. CH_2O/cat.amt. Et_2NH

Other reactions can follow in the wake of Michael reactions, as in the synthesis of dimedone, which shall be our finale.

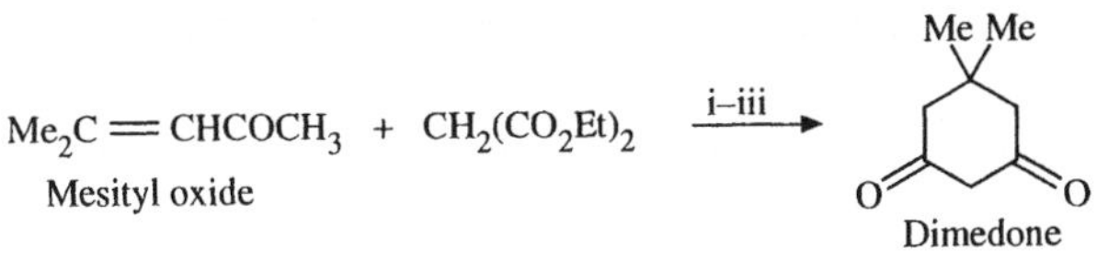

$$Me_2C = CHCOCH_3 \ + \ CH_2(CO_2Et)_2 \quad \xrightarrow{\text{i–iii}}$$

Mesityl oxide

Dimedone

Conditions: i, NaOEt/EtOH/Δ; ii, aq. KOH/Δ; iii, H$_3$O$^+$/Δ

This synthesis involves a Michael addition of one reactant to the other, followed by cyclization to give a strain-free six-membered ring; after saponification, the diketo-acid easily decarboxylates to give a dimedone enol.

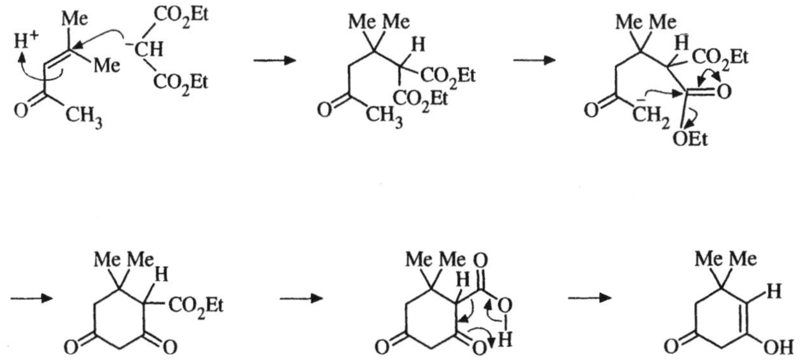

Problems

Explain the following:

1.

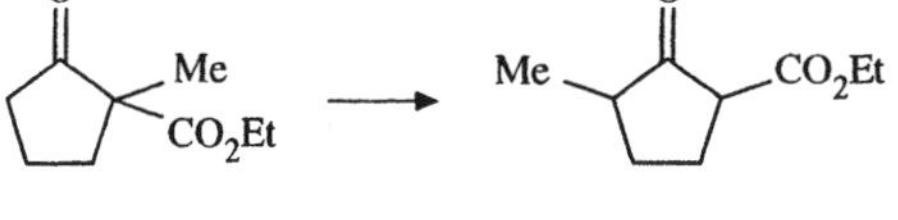

Conditions: i, NaOEt/EtOH; ii, H$_3$O$^+$ work-up

2.

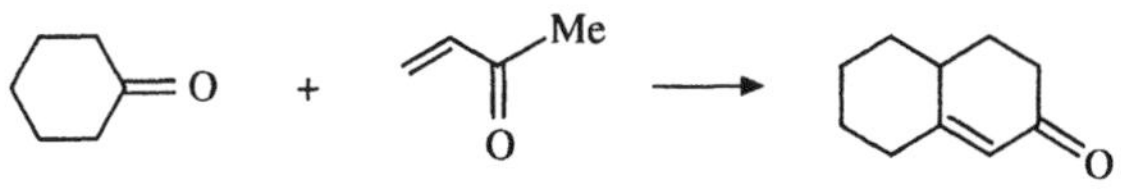

Conditions: i, NaOEt/EtOH; ii, H$_3$O$^+$ work-up

3.

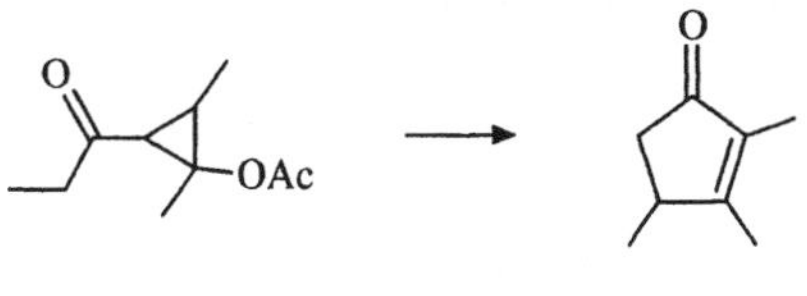

Conditions: i, NaOH/H$_2$O; ii, H$_3$O$^+$ work-up

4.

Conditions: H_3O^+

Epilogue

Even within the limited syllabus which was defined at the outset of Chapter 1, there are many variations on, and combinations of, the main themes of carbonyl chemistry we have outlined, about which nothing has been said. Reactions of enolates at the α-position with Se-electrophiles, for example; and at the carbonyl carbon with C-nucleophiles such as diazoalkanes, or with α-anions derived from nitroalkanes, sulfoxides, and sulfones, which are enolate analogues. Many of the chemical procedures for heteroaromatic synthesis depend on additions to and/or substitutions at carbonyl carbon (see Primer 2). And there are diverse lesser themes within the chemistry of aldehydes, ketones, and carboxylic acid derivatives which might have been touched on if space had allowed. These include reactions in which the key events involve molecular rearrangement (see Primer 5), the addition of an electron, excitation by light, the intermediacy of acyloxy-radicals, or cycloaddition, e.g. with alkylidenephosphoranes (Wittig reaction).

$$R \overset{O}{\underset{}{\big\|}} N - X \longrightarrow RN{=}C{=}O$$

Involved in the Curtius, Hofmann and Lossen rearrangements; cf. the Wolff rearrangement of diazoketones (Arndt–Eistert homologation)

$$\underset{R(OR)}{\overset{R}{\big\rangle}}{=}O \xrightarrow{1e} \underset{R(OR)}{\overset{R}{\big\rangle}}{-}O^-$$

Involved in the acyloin synthesis and pinacol coupling

$$\underset{R}{\overset{R}{\big\rangle}}{=}O \xrightarrow{h\upsilon} \underset{R}{\overset{R}{\big\rangle}}{-}O^{\cdot}$$

Involved in the Norrish Type I and Norrish Type II photochemical reactions of carbonyl compounds

$$R \overset{O}{\underset{O^{\cdot}}{\big\|}} \longrightarrow R^{\cdot}$$

Involved in the Kolbe electrolysis and the Hunsdiecker reaction

$$\underset{R}{\overset{R}{\big\rangle}}{=}O \; + \underset{R}{\overset{R}{\big\rangle}}{-}PPh_3 \longrightarrow \underset{R}{\overset{R}{\big\langle}}\overset{O}{\underset{PPh_3}{\big\rceil}}\underset{R}{\overset{R}{\big\rangle} \longrightarrow \underset{R}{\overset{R}{\big\rangle}}{=}\underset{R}{\overset{R}{\big\langle}} \quad \text{Wittig reaction}$$

Notes on the end of chapter problems

Chapter 1

1. Reaction at oxygen can occur as shown below. Methyl iodide, however, is a relatively soft electrophile, and reacts preferentially at carbon with simple enolates, as that is the softer end.

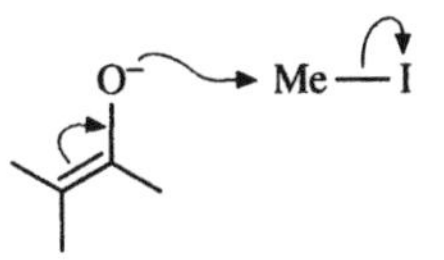

2. A reaction pathway is open to acetyl chloride in which negative charge is pushed onto oxygen when the ethoxide anion attacks, giving a reasonably stabilized tetrahedral intermediate. The corresponding mechanism is not possible with 2-chloropropene, because it would push charge onto a carbon which is out on a limb, without any electron-accepting substituent attached.

3. Diazomethane reacts as $^-CH_2$-N_2^+, attacking the electrophilic carbon of the ketone giving a tetrahedral adduct, which can collapse in two ways to the respective products.

Chapter 2

1. The Grignard reagent opens the hemiacetal ring, revealing an aldehyde function which reacts with a further mole of Grignard reagent in the usual way.

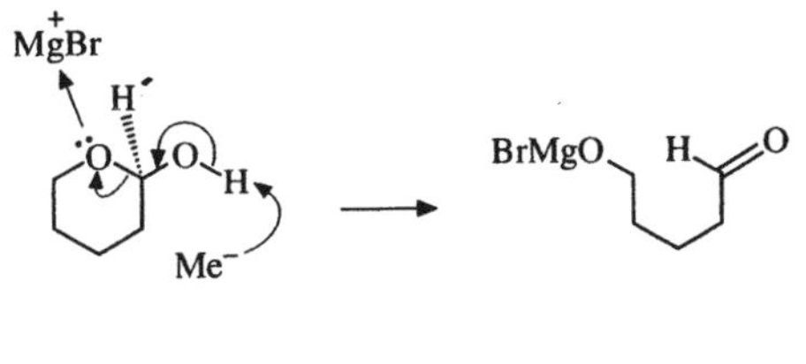

2. This is an industrial process for the preparation of methyl methacrylate, an important polymer precursor. Reaction takes place through the cyanohydrin, which undergoes dehydration to give a C=C bond, and methanolysis of the nitrile group.

3. The Grignard reagent reacts in the usual way to give a tertiary alcohol.

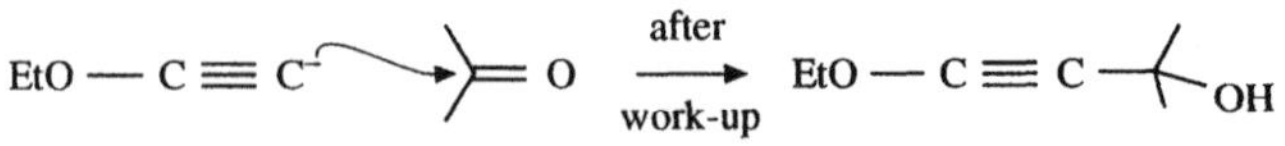

The OH group is protonated during workup, yielding after loss of water a delocalized cation.

The delocalized cation reacts with water at * to give an enol which promptly tautomerizes to the observed product.

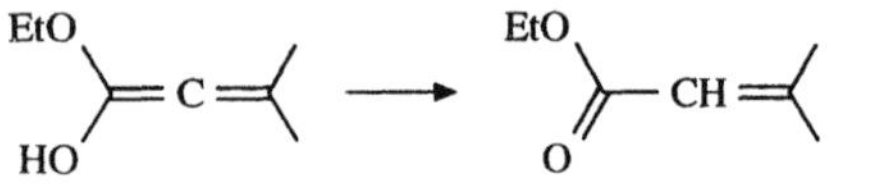

Chapter 3

1. See Primer 33, p. 19.

2. The Grignard reagent reacts with acetone in the usual way to give a tertiary alcohol; aqueous acid reveals an aldehyde function which is trapped by the OH group as a hemiacetal; oxidation of the hemiacetal secondary OH group gives the observed lactone.

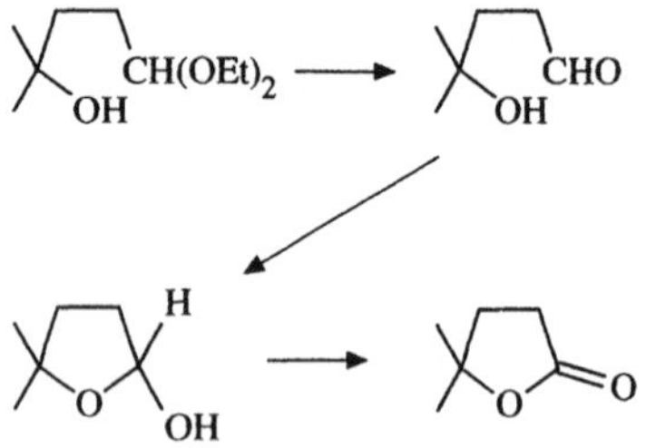

Chapter 5

1. A hydrazone is formed by reaction at the ketone carbonyl and the remaining amino group then attacks the ester carbonyl, giving a pyrazolone.

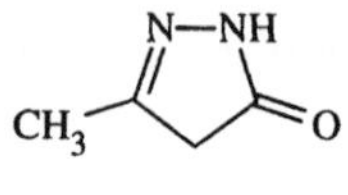

2. The other product is benzyl bromide, formed by an $S_N2(A_{AL}2)$-like attack of bromide ion on the protonated ester.

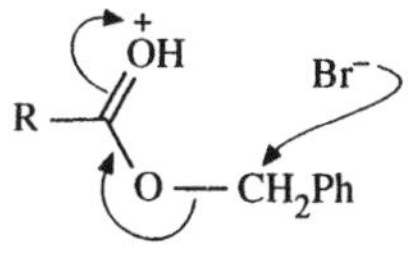

3. The mechanism here is also $S_N2(A_{AL}2)$-like; water attacks the protonated lactone on the alkyl side of the ester function, with inversion of configuration at the reaction site, a general feature of S_N2 reactions.

Chapter 6

1. A mixed anhydride of the starting acid and pivalic acid is formed; this reacts preferentially with benzylamine at the carbonyl group from the original acid, because the other one is sterically hindered. This is one of many procedures for activating carboxyl groups.

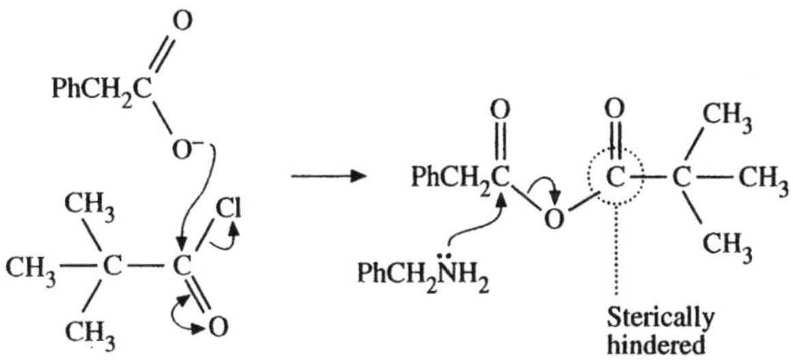

2. The bond between the CO and N of the four-membered β-lactam ring. The N is constrained by the fused ring geometry, so that its three bonds cannot all lie coplanar with each other and the CO group, as is required for the conjugation which deactivates normal amide bonds to nucleophilic attack at the CO group. Furthermore, the β-lactam ring is strained, and the system is like a compressed spring, ready to fly open.

Chapter 7

1. In water, solvent competes for the hydrogen bonds which stabilize the enol (see Section 7.1, in the margin).

2. There are many ways of achieving this conversion. The simplest is α-bromination and displacement of bromide with ammonia as below. See Primer 7, Chapter 1, for more possibilities.

$$CH_3CO_2H \longrightarrow BrCH_2CO_2H \longrightarrow NH_2CH_2CO_2H$$

3. The most acidic hydrogens are as follows.

In the case of methyl propyl ketone, the most stable enolate is the one which has most substitutents on the C=C bond, so the CH_2 group next to the CO is more acidic than the methyl group on the other side of it; the remaining hydrogens are of course not acidic at all. In the case of the $\alpha\beta$-unsaturated ketone, ionization by loss of a proton from the methyl group most distant from the CO group gives a more delocalized anion than ionization from any other position.

Chapter 8

1. It is impossible to place a double bond at a bridgehead, because it would be so twisted that there would be no π-overlap of the necessary p-orbitals, which would in fact be orthogonal (try making a model!). This is Bredt's rule. Because the enolate which would be generated by ionization of the bridgehead hydrogen would violate Bredt's rule, it is not formed. Only the hydrogens of the CH_2 next to the CO can be removed to give an enolate. A single enolate is formed—it does not matter which of the two hydrogens is lost to the base—but methyl iodide reacts preferentially on the least hindered face to give a single product.

2. Enolates also react with alkylating agents such as epoxides. In this example lactone formation ensues; hydrolysis and decarboxylation in the usual way gives γ-butyrolactone. If the lactone ring opened during hydrolysis and decarboxylation it would close again on isolation, as such lactone rings are formed with great ease.

3. This coupling reaction involves successive iodination and alkylation, giving a tetraester which undergoes hydrolysis and decarboxylation at both ends in the usual way to give succinic acid.

Chapter 9

1. It is tempting to propose that what is involved here is nucleophilic attack across the ring as shown below.

But this is stereoelectronically impossible. The displacement of an enolate from a methyl group as shown would be improbable in any case, because only good leaving groups are so displaced by nucleophiles, but there is also a fundamental objection to the proposal. For an S_N2-type displacement, the nucleophile must be able to get behind the reaction site and push out the leaving group in line on the other side, which it cannot do here. What actually happens is that the ring is opened by a reverse Dieckmann reaction to give a diester which then cyclizes the other way round. Isomerisation takes place in the direction shown because the product is a much stronger acid than the starting ester; in the product there is a CH which lies between two CO groups, whereas in the starting ester the ionizable hydrogens are only activated by one CO group.

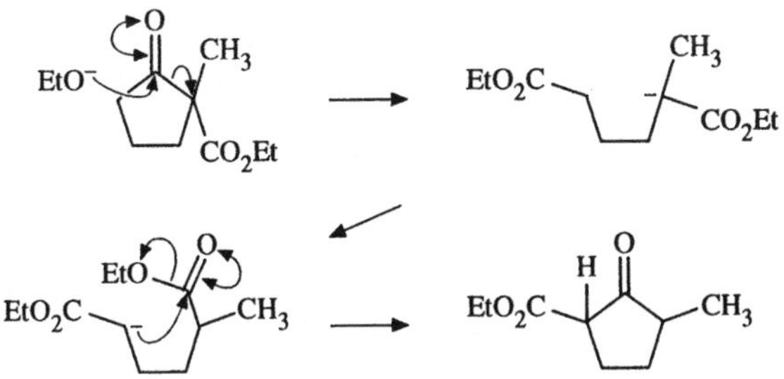

2. This is an example of the Robinson ring extension. It involves successive Michael addition to give A, aldol cyclization to give B and finally dehydration to give the observed product C.

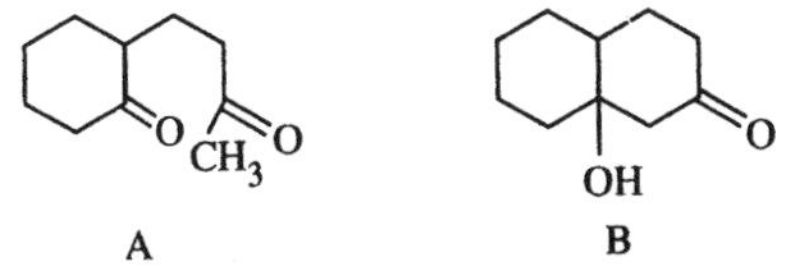

3. $B_{AC}2$ ester hydrolysis is followed by a reverse aldol reaction, which is driven by release of ring strain. This gives an open-chain γ-diketone, which cyclizes by a forward aldol reaction to give a comfortable 5-ring, followed by loss of water to give the observed product. Actually the cyclization could have taken place to give another 5-ring product: reflect on this.

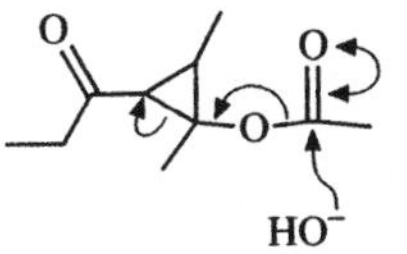

4.   This is a variant of Robinson's tropinone synthesis, in which a naturally occurring complex ring system was prepared under very mild conditions from rather simple precursors, mimicking the biosynthetic process. The dimethoxytetrahydrofuran starting material is a double acetal and is easily hydrolysed under acidic conditions to succindialdehyde. Two Mannich reactions follow, and the synthesis is completed by loss of the two carboxyl groups, both of which are β to a ketone carbonyl.

Additional reading

The chemistry of carbonyl compounds is so central to organic chemistry that several of the Primers overlap the topic and treatments are found in all textbooks. At the present time students are spoilt for choice, but perhaps the relevant chapters of McMurry (there is an excellent brief introductory account as well as the full textbook version), Sykes, and Norman and Coxon will be found most useful. Kemp and Vellaccio is a model of clarity. For a programmed introduction (a step-by-step problem-solving approach, that is) Warren's book has stood the test of time. Many reactions of carbonyl compounds generate new chiral centres, so asymmetric synthesis is an important aspect of their chemistry: it is well covered by Procter. On points of biological relevance, see Abeles, Frey, and Jencks or Stryer.

Abeles, R. H., Frey, P. A., and Jencks, W. P. (1992). *Biochemistry*. Jones and Bartlett, London.

Kemp, D. S. and Vellaccio, F. (1980). *Organic chemistry*. Worth, New York.

McMurry, J. (1994). *Fundamentals of organic chemistry*, (3rd edn). Brooks Cole, California. Especially Chapters 9–11.

McMurry, J. (1995). *Organic chemistry*, (4th edn). Brooks Cole, California. Especially Chapters 19–23.

Procter, G. (1996). *Asymmetric synthesis*. Oxford University Press, Oxford.

Norman, R. O. C. and Coxon, J. M. (1993). *Principles of organic synthesis* (3rd edn). Blackie, London.

Stryer, L. (1996). *Biochemistry* (4th edn). W. H. Freeman, New York.

Sykes, P. (1986). *A guidebook to mechanism in organic chemistry* (6th edn). Longman, London. See also Sykes, P. (1995). *A primer to mechanism in organic chemistry*. Longman, London. Especially Chapter 7.

Warren S. (1974, but reprinted frequently and still in print 1997). *Chemistry of the carbonyl group*. Wiley, London.

Index

A detailed contents breakdown is given at the beginning of the book as an additional finding aid. Major topics and classes of compounds which can easily be located by using it are not indexed.